W0080947

Harish-Chandra Research Institute
Lecture Notes - 1

Arithmetical Aspects of the Large Sieve Inequality

Arithmetical Aspects of the Large Sieve Inequality

Olivier Ramaré

CNRS, Université Lille 1
France

With the Collaboration of

D. S. Ramana
Harish-Chandra Research Institute
Allahabad, India

 HINDUSTAN
BOOK AGENCY

Published by

Hindustan Book Agency (India)
P 19 Green Park Extension
New Delhi 110 016
India

email: hba@vsnl.com
http://www.hindbook.com

Copyright © 2009 by Hindustan Book Agency (India)

No part of the material protected by this copyright notice may be
reproduced or utilized in any form or by any means, electronic or
mechanical, including photocopying, recording or by any information storage and retrieval system, without written permission from
the copyright owner, who has also the sole right to grant licences
for translation into other languages and publication thereof.

All export rights for this edition vest exclusively with Hindustan
Book Agency (India). Unauthorized export is a violation of
Copyright Law and is subject to legal action.

Produced from camera ready copy supplied by the Authors.

ISBN-13 978-81-85931-90-6

Preface

These lectures were given in February 2005 while I was a guest of the Harish-Chandra Research Institute, and the bulk of these notes was written while I was staying there.

Though this course was intended for people having some background in analytic number theory, efforts have been made to restrict the prerequisites to a minimum. As an effect, most of these notes can be read with no prior knowledge in the area, except for some applications which require the prime number Theorem for arithmetic progressions and, in some places, the Bombieri-Vinogradov Theorem.

I wish to thank the Harish-Chandra Research Institute for giving me the opportunity to give this series of lectures and for providing extremely agreeable surroundings, the CEFIPRA programme "Analytic and Combinatorial Number Theory", project 2801-1 directed by Professors Bhowmik and Balasubramanian, for funding most of my journey, and finally my host, Professor Adhikari, without whom none of this would have been possible. I am also indebted to S. Baier who attended these lectures and pointed out useful references, as well as to the other persons in the audience for questions that helped me clarify these notes.

Professor D. Surya Ramana has been of great help during the writing of this monograph: he has read many a new version, checked formulae, corrected references as well as provided a most welcomed linguistic support. Chapter 3 is his, so is the last part of section 1.2.1 as well as several parts of the proofs presented. Both of us would like to thank the Indo-French Institute for Mathematics for supporting this collaboration.

O. Ramaré

Contents

Introduction

The idea of the large sieve appeared for the first time in the foundational paper of (Linnik, 1941). Later (Rényi, 1950), (Barban, 1964), (Roth, 1965), (Bombieri, 1965), (Davenport & Halberstam, 1966b) developed it and in particular, two distinct parts emerged from these works:

(1) An analytic inequality for the values over a well-spaced set of points of a trigonometric polynomial $S(\alpha) = \sum_{1 \leq n \leq N} u_n e(n\alpha)$, which, in arithmetical situations, most often reduces to

(0.1) $$\sum_{q \leq Q} \sum_{a \bmod^* q} \left|S(a/q)\right|^2 \leq \Delta \sum_n |u_n|^2$$

for some Δ depending on the length N of the trigonometric polynomial and on Q. The best value in a general context is $\Delta = N - 1 + Q^2$ obtained independently in (Selberg, 1972) and in (Montgomery & Vaughan, 1973).

(2) An arithmetical interpretation for $\sum_{a \bmod^* q} \left|S(a/q)\right|^2$, where this time, information on the distribution of (u_n) modulo q is introduced. The most popular approach goes through a lower bound and is due to Montgomery, leading to what is sometimes referred to as *Montgomery's sieve*, by reference to (Montgomery, 1968).

Today the terminology *large sieve* refers to a combination of the two aforementioned steps. We refer the reader to the excellent lecture notes (Montgomery, 1971) and the survey paper (Montgomery, 1978) for the early part of the development, but cite here the papers of (Bombieri & Davenport, 1968) and (Bombieri, 1971).

Almost simultaneously, (Selberg, 1949) introduced another way of sieving, which we now describe rapidly in the following simple form for the primes: to find an upper bound for the number of primes in the interval $]\sqrt{N}, N]$, consider the following inequality

(0.2) $$\sum_{\sqrt{N} < p \leq N} 1 \leq \sum_{n \leq N} \left(\sum_{d \mid n} \lambda_d\right)^2$$

valid for any λ_d's subject to $\lambda_1 = 1$ and $\lambda_d = 0$ if $d > z$ for some parameter $z \leq \sqrt{N}$. This leads to the determination of the minimum of the quadratic form on the R.H.S. of (0.2), a method for which Selberg designed an appropriate elementary method.

The similarity between the large sieve procedure and Selberg's is far from obvious, but one readily notes that both of these are based on an L^2-kind of argument, and that both rely on an arithmetical inequality which is controlled only extremely loosely. Moreover it turns out that both, despite their simplicity, lead to best results in sieve theory (provided the sieve dimension is ≥ 1).

That both of these procedures were related became apparent at least in the early seventies as can be seen from the papers of (Huxley, 1972b), (Kobayashi, 1973) and (Motohashi, 1977), so word went around that both sieves are dual to each other, at least in a vague sense, though the papers quoted above of course give a precise meaning to this suggested duality. Things get somewhat more intricate if one notices that the large sieve inequality may be proved via its dual form as in (Elliott, 1971). Let us mention here that this very flexible process usually leads to bounds of good quality.

Our aim in these lectures is to develop a unique setting for the large sieve and Selberg sieve, based on hermitian inequalities. This can be seen as an elaboration of ideas due to Selberg, as exposed in (Bombieri, 1987). Along the way, we shall meet, recognize and show links between notions used at different places.

In the first stage, we extend the classical arithmetic form of the large sieve, in a fashion very much inspired by (Bombieri & Davenport, 1968). This generalization will have consequences, and we shall in particular improve on the large sieve inequality when applied to *sifted sequences*. Our closer scrutiny will provide a *large sieve extension* of the sieve bound but only under a specific condition, thus showing some discrepancy between both sieving processes, besides the fact that the large sieve applies only when sieving intervals while Selberg sieve encompasses the case of general sequences. By a *large sieve extension*, we mean that we are able to bound not only the number of points satisfying some congruence conditions, but also are able to give an upper bound for quantities measuring distribution in arithmetic progressions, as in the theorems of (Barban, 1963), (Barban, 1964), (Barban, 1966), (Davenport & Halberstam, 1966a) and (Davenport & Halberstam, 1968).

In the second stage, we develop a theory of what we call *local models*, essentially through examples. Roughly speaking, we build an approximation of the function we are interested in modulo q by multiplying a model for its reduction modulo q and a model for its behavior from the point of view of size condition (our place at infinity, to use the language of number theory). As an application we shall prove a large sieve type inequality but with an error term similar to the one appearing in Selberg sieve and improve on the asymptotic Brun-Titchmarsh inequality. This third approach will show how the two previous ones, via the large sieve and via Selberg sieve, are connected. But it will also lead to further developments and, in particular, to some results on some binary additive problems, via a method not unlike an abstract circle method. We note here that (Heath-Brown, 1985) has already pointed in this direction.

In the third stage, drawing on what we introduced earlier, we present the Selberg sieve in an elementary fashion so as to encompass the case of non-squarefree sifting conditions. This approach will apply to sequences as well, while earlier expositions in (Selberg, 1976) or (Motohashi, 1983) did not. Moreover we shall also understand Selberg's pseudo-characters (see for instance (Motohashi, 1983) for a definition) and extend the result of (Kobayashi, 1973) to our more general situation. This part will also show links between this sieving process and approximation of the van Mangoldt function Λ as it appears, for instance, in (Motohashi, 1978), (Heath-Brown, 1985), (Goldston, 1992) or (Iwaniec, 1994). As a matter of fact, this line of thought arose from ideas at the very origin of Selberg sieve, see (Selberg, 1942).

In the fourth and final stage, we develop our material in several directions. We first show the classical theorem of (Bombieri & Davenport, 1966) on prime gaps by our method, and in particular without any use of the circle method. We also handle in a similar fashion the case of the representation of an integer by a sum of two squarefree numbers. It is at this that we shall prove a general approximation theorem for a function by local models: we delayed such a statement this much because it requires a clarification of the notion of local model, notably concerning the way to handle the infinite place. We end our journey by discussing which binary problems are accessible through this pass, meeting here with some material due to (Brüdern, 2000-2004) and some due to (Friedlander & Iwaniec, 1992).In between, we shall expand on the particularly elegant smoothing functions due to (Holt & Vaaler, 1996) that will allow us to prove a novel generalization of the large sieve inequality, while simplifying estimations in the context of our local models.

We have attempted to present all this material in a manner as elementary as possible, and this sometimes prevents us from gaining some height. Already as such, we require several unusual definitions. For this reason we have supplemented our exposition with the chapter 4 and 14, which describe with greater care the surroundings and prepare the ground for a more axiomatic approach. In particular, we insist on getting what we call a *geometrical* interpretation to connect our combinatorial constructions with properties of sets such as $\mathbb{Z}/d\mathbb{Z}$. The situation

is more difficult than that, and indeed, eventually, we will contend with
properties on the space of functions on such sets.

We finally mention that Motohashi has developed the arithmetical
setting of the large sieve in a very different direction, see for instance (Motohashi, 1983). Moreover, many arithmetical applications of the large
sieve inequality stem from its multiplicative form, a subject which we
shall not touch upon: the reader is referred to the excellent broach
of (Bombieri, 1987). Among general references on the subject, we mention the books of (Halberstam & Richert, 1974) and of (Huxley, 1972a).
Furthermore, Elsholtz has developped combinatorial uses of the large
sieve inequality, a subject we shall not touch at all; We simply refer
to (Elsholtz, 2001), (Elsholtz, 2002), (Elsholtz, 2004) and (Croot III &
Elsholtz, 2004). Finally, the reader will find in (Huxley, 1968), (Huxley,
1970) and (Huxley, 1971) material pertaining to a large sieve inequality
for algebraic number fields as well as several applications of it.

Individual chapters in these notes are meant to present a circle of
ideas, with references given therein to other parts where a different point
of view is taken, or where one has an easier access to certain lemmas or
notions. Such a choice is rendered necessary by the subject itself: we
intend showing different developments in a unified context, but these
developments are in fact quite entangled one with another. We study
several examples, some of them leading to new results, but limited some
of the proofs to illuminating special cases.

A final word on averages of non-negative multiplicative functions.
Evaluating such averages is a most commonly met question, and we
have decided to present the convolution method as well as a number of
results originating from (Levin & Fainleib, 1967). We have isolated the
main result of this celebrated paper in an appendix, in a slightly more
general form required in our context and took the opportunity to detail
there two classical examples. However, since these results are scattered
throughout the monograph, here is an index:

(1) Lemma 2.3 is a generalization of a lemma due to (van Lint &
    Richert, 1965).
(2) Proof of Theorem 2.2, page 23: an ad-hoc lower bound.
(3) Section 5.3, page 42 starts with a sketch of the convolution
    method.
(4) Proof of Theorem 5.4 contains page 45 another example on the
    convolution method.
(5) Proof of Lemma 6.2, page 57 relies on the idea of (Levin &
    Fainleib, 1967).

(6) Theorem 9.2, page 77 is yet another use of this idea.

(7) Section 13.3, page 112 contains an application of our version of the Levin-Fainleib Theorem, namely Theorem 21.1, while section 13.5 contains another one.

(8) The appendix presents statement and proof of this Theorem 21.1, together with yet another instance of its use.

The reader should however be aware that the theory is in no way restricted to these two lines of approach and will consult with benefit (Wirsing, 1961), (Halász, 1971/72), (Montgomery & Vaughan, 2001) and (Granville & Soundararajan, 2003).

Multiplicativity and its numerous variations are freely used throughout this book, as is the arithmetical convolution. We have tried to stick to common notations and to summarise most of them page 187. We hope that this summary, together with the reference index, will help the reader navigate at his or her own will within this monograph!

1 The large sieve inequality

We begin with an abstract hermitian setting which we will use to prove the large sieve inequality. We develop more material than is required for such a task. This is simply to prepare the ground for future uses, and we shall even expand on this setting in chapter 7; the final stroke will only appear in section 10.1.

1.1. Hilbertian inequalities

Let us start with a complex vector space $\mathcal{H}$ endowed with a hermitian form $[f|g]$, left linear and right sesquilinear. To be consistent with later notations, the norm of φ is denoted by $\|\varphi\|_2$.

The easiest exposition goes through a formal definition:

Definition 1.1. *By an* almost orthogonal system *in* $\mathcal{H}$, *we mean a collection of three sets of data*

    (1) a finite family $(\varphi_i^*)_{i \in I}$ *of elements[1] of* $\mathcal{H}$,
    (2) a finite family $(M_i)_{i \in I}$ *positive real numbers,*
    (3) a finite family $(\omega_{i,j})_{i,j \in I}$ *of complex numbers with* $\omega_{j,i} = \overline{\omega_{i,j}}$,

all of them given so that

$$(1.1) \qquad \forall (\xi_i)_i \in \mathbb{C}^I, \quad \left\|\sum_i \xi_i \varphi_i^*\right\|_2^2 \leq \sum_i M_i |\xi_i|^2 + \sum_{i,j} \xi_i \overline{\xi_j} \omega_{i,j}.$$

We comment on this definition. If the family $(\varphi_i^*)_{i \in I}$ were orthogonal, we could ask for equality with $M_i = \|\varphi_i^*\|_2^2$. As it turns out, in the applications we have in mind, this family is not orthogonal, but almost so. It is this almost orthogonality that the above condition is meant to measure.

Our first lemma reads as follows

Lemma 1.1. *For any finite family* $(\varphi_i^*)_{i \in I}$ *of points of* $\mathcal{H}$, *the system built with* $M_i = \sum_j |[\varphi_i^*|\varphi_j^*]|$ *and* $\omega_{i,j} = 0$ *is almost orthogonal.*

So that if $[\varphi_i^*|\varphi_j^*]$ is small for $i \neq j$ then M_i is indeed close to $\|\varphi_i^*\|_2^2$

---

[1] The reader may wonder why I chose to denote the members of this family with a star ... It is to be consistent and to avoid confusion with notation that will appear later on.

Proof. We write

$$\left\| \sum_i \xi_i \varphi_i^* \right\|_2^2 = \sum_{i,j} \xi_i \overline{\xi_j} [\varphi_i^* | \varphi_j^*]$$

and simply apply $2|\xi_i \overline{\xi_j}| \leq |\xi_i|^2 + |\xi_j|^2$. The lemma readily follows. $\diamond\diamond\diamond$

Here is an enlightening reading of this lemma: the hermitian form that appears has a matrix whose diagonal terms are the $\|\varphi_i^*\|_2^2$'s. A theorem of Gershgorin says that all the eigenvalues of this matrix lie in the union of the so called *Gershgorin's discs* centered at the points $\|\varphi_i^*\|_2^2$, with radius $\sum_{j \neq i} |[\varphi_i^* | \varphi_j^*]|$. This approach is due to (Elliott, 1971). It has a drawback: we do not know that each Gershgorin disc does indeed contain an eigenvalue, a flaw that is somehow repaired in the above lemma.

In general, and only assuming (1.1), we get the following kind of Parseval inequality:

Lemma 1.2. *For any almost orthogonal system, and any $f \in \mathcal{H}$, let us set $\xi_i = [f|\varphi_i^*]/M_i$. We have*

$$\sum_i M_i^{-1} |[f|\varphi_i^*]|^2 \leq \|f\|_2^2 + \sum_{i,j} \xi_i \overline{\xi_j} \omega_{i,j}.$$

Once again, the orthogonal case is enlightening: if the (φ_i^*) are orthogonal, then we may take $M_i = \|\varphi_i^*\|_2^2$ and $\omega_{i,j} = 0$. The L.H.S. becomes the square of the norm of the orthonormal projection of f on the subspace generated by the φ_i^*'s.

Without the $\omega_{i,j}$ and appealing to Lemma 1.1, this is due to Selberg, as mentioned in section 2 of (Bombieri, 1987) and in (Bombieri, 1971).

Proof. For the proof, we simply write

$$\left\| f - \sum_i \xi_i \varphi_i^* \right\|_2^2 \geq 0$$

and expand the square. We take care of $\|\sum_i \xi_i \varphi_i^*\|_2^2$ by using (1.1), getting

$$\|f\|_2^2 - 2\Re \sum_i \overline{\xi_i} [f|\varphi_i^*] + \sum_i M_i |\xi_i|^2 + \sum_{i,j} \xi_i \overline{\xi_j} \omega_{i,j} \geq 0.$$

We now choose the ξ_i's optimally, neglecting the bilinear form containing the $\omega_{i,j}$. We take $\xi_i = [f|\varphi_i^*]/M_i$, the lemma readily follows. $\diamond\diamond\diamond$

Combining Lemma 1.2 together with Lemma 1.1 yields what is usually known as "Selberg's lemma" in this context. The introduction of the $\omega_{i,j}$ is due to the author to enable a refined treatment of the error term as well as provide a hybrid between the weighted large sieve results and Selberg sieve results. In these lectures however, we shall only have a glimpse of this aspect. Nevertheless we show in chapter 9 a simpleminded use of this bilinear part.

The actual value of ξ_i in the statement is usually of no importance, only its order of magnitude being relevant.

Let us end this section with a historical remark: though the material presented here is recent, the reader will find in the seventh part of (Rényi, 1958) a similar approach, relying on the notion, borrowed from (Boas, 1941), of *quasi-orthogonal sequence of random variables*. Furthermore, (Rényi, 1949) already introduces a notion of quasi-orthogonality in the context of the large sieve inequality. We close this parenthesis and refer the reader to (Montgomery, 1971) for more historical material.

1.2. The large sieve inequality

The large sieve inequality reads as follows.

Theorem 1.1. *Let $\mathcal{X}$ be a finite set of points of $\mathbb{R}/\mathbb{Z}$. Set*

$$\delta = \min \left\{ \|x - x'\|, x \neq x' \in \mathcal{X} \right\}.$$

For any sequence of complex numbers $(u_n)_{1 \leq n \leq N}$, we have

$$\sum_{x \in \mathcal{X}} \left| \sum_n u_n e(nx) \right|^2 \leq \sum_n |u_n|^2 (N - 1 + \delta^{-1}).$$

The L.H.S. can be thought as a Riemann sum over the points in $\mathcal{X}$; at least when the set $\mathcal{X}$ is dense enough. The spacing between two consecutive points being at least δ, this L.H.S. multiplied by δ can thought as approximating

$$\int_0^1 \left| \sum_n u_n e(n\alpha) \right|^2 d\alpha = \sum_n |u_n|^2.$$

This is essentially so if δ^{-1} is much greater than N, but it turns out that the case of interest in number theory is the opposite one. In this case, we can look at $\sum_n u_n e(nx)$ as being a linear form in $(u_n)_n$. The spacing condition implies that $\mathcal{X}$ has less than δ^{-1} elements, so that the number of linear forms implied is indeed less than the dimension of the ambient space (which is N). In that case these linear forms are independent as shown by computing a van der Monde determinant, and otherwise, there is some redundancy. So what is really at stake here is more almost orthogonality than approximation, which is why I chose this method of proof.

The theorem in this version is due to Selberg. The same year and by a different method, a marginally weaker version (without the -1 on the right) was proved by (Montgomery & Vaughan, 1973). We shall prove a slightly weaker result, namely with $N+1+2\delta^{-1}$ instead of $N-1+\delta^{-1}$ in this chapter and delay a full proof until chapter 15, where we shall also provide a generalization. First we recall what is the Fourier transform of the de la Vallée-Poussin kernel.

1.2.1. A Fourier transform.

Let N' and L be two given positive integers. Consider the function $F(n)$ whose graph is:

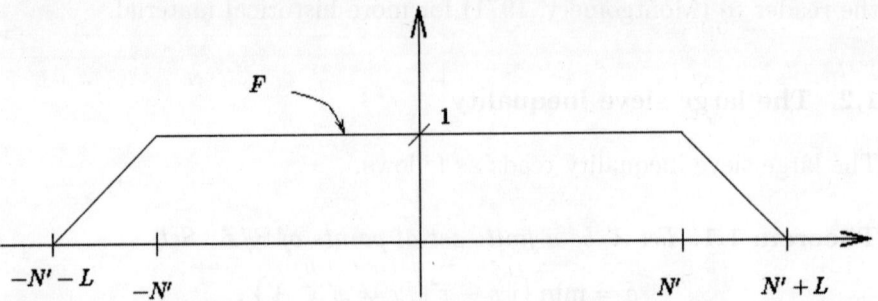

We are to compute its Fourier transform which can be cumbersome. We present two proofs, the first one being more geometrical but only adapted to the present situation while the second one is less visual but often trivialises computations of this kind.

First proof. To simplify calculations, we write $F = (G-H)/L$ where G and H are drawn below. We write

$$L\sum_{n\in\mathbb{Z}}F(n)e(ny) = \sum_{n\in\mathbb{Z}}G(n)e(ny) - \sum_{n\in\mathbb{Z}}H(n)e(ny)$$

$$= \sum_{0\leq|n|\leq N'+L}(N'+L-|n|)e(ny) - \sum_{0\leq|n|\leq N'}(N'-|n|)e(ny)$$

and obtain

$$L\sum_{n\in\mathbb{Z}}F(n)e(ny) = \left|\sum_{0\leq m\leq N'+L}e(my)\right|^2 - \left|\sum_{0\leq m\leq N'}e(my)\right|^2.$$

This finally amounts to

$$(1.2)\qquad \sum_{n\in\mathbb{Z}}F(n)e(ny) = \frac{1}{L}\left|\frac{\sin\pi(N'+L)y}{\sin\pi y}\right|^2 - \frac{1}{L}\left|\frac{\sin\pi N'y}{\sin\pi y}\right|^2,$$

the value at $y=0$ being given by $\sum_{n\in\mathbb{Z}}F(n) = 2N'+L$.

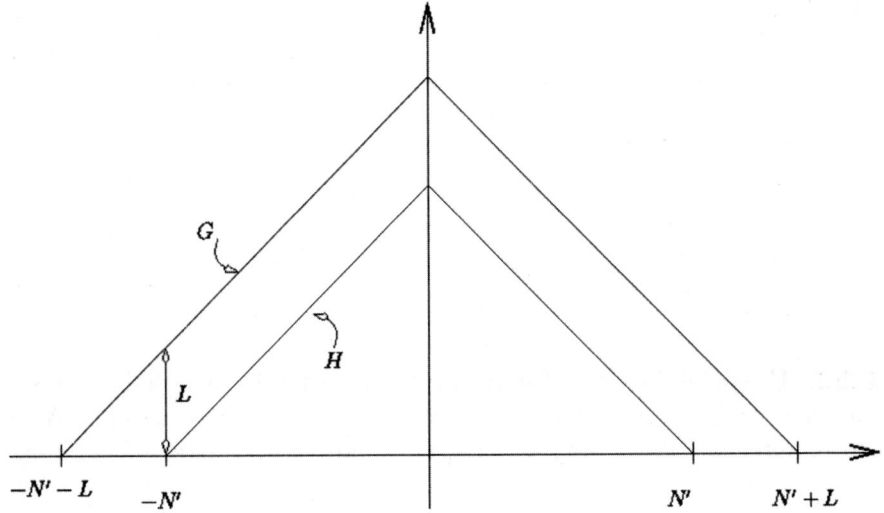

Second proof. Let us define

$$(1.3) \qquad f(y) = \sum_{n \in \mathbb{Z}} F(n) e(ny)$$

and introduce the operator on compactly supported sequences:

$$(1.4) \qquad \Delta(F) = \Delta\big((F(n))_{n \in \mathbb{Z}}\big) = \big((F(n) - F(n-1))_{n \in \mathbb{Z}}\big).$$

We readily see that

$$\begin{aligned}
f(y)(e(y) - 1) &= \sum_{n \in \mathbb{Z}} F(n) e(ny)(e(y) - 1) \\
&= -\sum_{n \in \mathbb{Z}} \Delta(F)(n)\, e(ny)
\end{aligned}$$

which is our main equation. Iterating once, we get

$$(1.5) \qquad f(y)(e(y) - 1)^2 = \sum_{n \in \mathbb{Z}} \Delta^2(F)(n)\, e(ny).$$

The reader will check that $\Delta^2(F)(n) = F(n) - 2F(n-1) + F(n-2)$ and from there derive

$$(1.6) \quad L\Delta^2(F) = \mathbb{1}_{n=-N'-L+1} - \mathbb{1}_{n=-N'+1} - \mathbb{1}_{n=N'+1} + \mathbb{1}_{n=N'+L+1}.$$

This finally yields

$$Lf(y) = \frac{e(y)\big(e(-(N'+L)y) - e(-N'y) - e(N'y) + e((N'+L)y)\big)}{(e(y)-1)^2}$$

$$= \frac{\cos(2\pi(N'+L)y) - \cos(2\pi N'y)}{-2\sin(\pi y)^2}$$

$$= \frac{\sin^2(\pi(N'+L)y) - \sin^2(\pi N'y)}{\sin(\pi y)^2}.$$

as required.

1.2.2. Proof of (a weak form of) Theorem 1.1. We use Lemma 1.2 together with Lemma 1.1. First notice that we may assume N to be an integer. Next set $N' = \lfloor N/2 \rfloor$ the integer part of $N/2$ and $f(n) = u_{N'+1+n}$ (with $u_{N+1} = 0$ if N is even) so that f is supported on $[-N', N']$. The Hilbert space we take is $\ell^2(\mathbb{Z})$ with its standard scalar product so that f belongs to it when extended by setting $f(n) = 0$ for any integer n not in the interval above. Notice also that

$$(1.7) \qquad\qquad \|f\|_2^2 = \sum_n |u_n|^2.$$

We need to define our almost orthogonal system. We take

$$(1.8) \qquad\qquad \forall x \in \mathcal{X}, \quad \varphi_x^*(n) = e(nx)\sqrt{F(n)},$$

where F is as defined in section 1.2.1. Since f vanishes outside $[-N', N']$, we find that

$$(1.9) \qquad\qquad [f|\varphi_x^*] = e(-(N'+1)x) \sum_{1 \le n \le N} u_n e(nx).$$

The computations of the preceding section show that

$$(1.10) \qquad \|\varphi_x^*\|_2^2 = 2N' + L, \quad |[\varphi_x^*|\varphi_{x'}^*]| \le \frac{1}{4L\|x-x'\|^2} \quad \text{if } x \ne x'$$

by using the classical inequality $|\sin x| \le 2\|x\|/\pi$. When x is fixed, we find that

$$\sum_{\substack{x' \in \mathcal{X} \\ x' \ne x}} |[\varphi_x^*|\varphi_{x'}^*]| \le \sum_{\substack{x' \in \mathcal{X} \\ x' \ne x}} \frac{1}{4L\|x-x'\|^2}$$

$$\le 2 \sum_{k \ge 1} \frac{1}{4L(k\delta)^2} \le \frac{\pi^2}{12L\delta^2}$$

since the definition of δ implies that the worst case that could happen for the sequence $(\|x - x'\|)_{x'}$ would be if all x''s were located at $x + \ell\delta$ with ℓ an integer taking the values $\pm 1, \pm 2, \pm 3, \ldots$. Next we choose L

an integer so as to nearly minimize $2N' + L + \pi^2/(12L\delta^2)$, i.e., with $\lceil x \rceil$ denoting the least integer larger than x,

(1.11)
$$L = \left\lceil \frac{\pi}{2\sqrt{3}\delta} \right\rceil$$

which yields $2N' + L + \pi^2/(12L\delta^2) \leq N + 1 + \frac{\pi}{\sqrt{3}}\delta^{-1}$. We conclude by noting that $\pi/\sqrt{3} \leq 1.82 \leq 2$.

Let us end this section by a methodological remark : (Montgomery, 1971) proves in an appendix the inequality $(\sin \pi x)^{-2} \leq (\pi\|x\|)^{-2} + 1$ valid for $0 \leq x \leq 1/2$. On using it we obtain a better bound for $[\varphi_x^* | \varphi_{x'}^*]$ above and consequently, improve our $N + 1 + \pi\delta^{-1}/\sqrt{3}$ to $N + 3 + 2\delta^{-1}/\sqrt{3}$.

1.3. Introducing Farey points

In most arithmetical applications, the set $\mathcal{X}$ is simply a truncation of the Farey series, that is

(1.12)
$$\mathcal{X} = \{a/q,\ q \leq Q, a \bmod^* q\}$$

where Q is a parameter to be chosen and $a \bmod^* q$ means a ranging over all the invertible residue classes modulo q. Next when a/q and a'/q' are two distinct points of $\mathcal{X}$, we have

(1.13)
$$\left| \frac{a}{q} - \frac{a'}{q'} \right| = \frac{|aq' - a'q|}{qq'} \geq \frac{1}{qq'} \geq Q^{-2}$$

since $aq' - a'q$ is an integer that is distinct from 0.[2] We set classically

(1.14)
$$S(x) = \sum_{1 \leq n \leq N} u_n e(na/q)$$

and get

(1.15)
$$\sum_{q \leq Q} \sum_{a \bmod^* q} |S(a/q)|^2 \leq \sum_n |u_n|^2 (N + Q^2)$$

which is essentially what is referred to as *the large sieve inequality*. In chapter 20, we shall provide some cases where we are able to compute an asymptotic for the L.H.S.. Moreover, but only for a restricted family of sequences, we shall even be able to do so with Q being larger than $\sqrt{N}$ – while the main term will still be of order of $N \sum_n |u_n|^2$ –, thus dramatically improving on this inequality.

---

[2] By discussing whether $q = q'$ or not, one can enlarge this bound to $1/(Q(Q-1))$.

1.4. A digression: dual form and double large sieve

The large sieve inequality bounds $\sum_{x\in\mathcal{X}}|S(x)|^2$. If we open one $S(x)$, we see that this quantity is also

$$(1.16) \qquad \sum_{n,x} u_n \overline{S(x)} e(nx)$$

which can now be considered a bilinear form in the two sets of variables $(u_n)_n$ and $(S(x))_x$, simply by forgetting how $S(x)$ is defined in terms of the u_n's. Such an expression has been considered in (Bombieri & Iwaniec, 1986) where they obtain a bound for it now known as the *double large sieve inequality* (see also (Selberg, 1991)). This bound is of similar strength as the one given by Theorem 1.1, up to a multiplicative constant, when applied to our situation. This line of ideas leads us – though historically, it is the reverse process that occured – to consider the so-called *dual form of the large sieve inequality*, which concerns the expression obtained simply by exchanging the variables n and x:

$$(1.17) \qquad \sum_n \left| \sum_{x\in\mathcal{X}} S(x) e(nx) \right|^2$$

where this time $(S(x))_x$ is any sequence of complex numbers. Proceeding as before but with the variable x, the above expression is also

$$(1.18) \qquad \sum_{n,x} \overline{S(x)} W(n) e(-nx) \quad \text{with} \quad W(n) = \sum_{y\in\mathcal{X}} S(y) e(ny)$$

to which we apply the Cauchy-Schwarz inequality in the x-variable to get

$$\left(\sum_n \left| \sum_{x\in\mathcal{X}} S(x) e(nx) \right|^2 \right)^2 \le \sum_x |S(x)|^2 \sum_x \left| \sum_n W(n) e(-nx) \right|^2.$$

Applying the usual large sieve inequality to the latter sum, we end up with the dual form of the large sieve inequality:

$$(1.19) \qquad \sum_n \left| \sum_{x\in\mathcal{X}} S(x) e(nx) \right|^2 \le \sum_x |S(x)|^2 (N - 1 + \delta^{-1}).$$

1.5. Maximal variant

We record here a maximal version of Theorem 1.1 whose proof is not yet completely satisfactory. This theorem is due to (Montgomery, 1981), improving on an earlier result of (Uchiyama, 1972).

Theorem 1.2. *There exist a constant $C > 0$ with the following property. Let $\mathcal{X}$ be a finite set of points of $\mathbb{R}/\mathbb{Z}$. Set*

$$\delta = \min\left\{ \|x - x'\|, x \neq x' \in \mathcal{X} \right\}.$$

For any sequence of complex numbers $(u_n)_{1 \leq n \leq N}$, we have

$$\sum_{x \in \mathcal{X}} \max_{K \leq N} \left| \sum_{1 \leq n \leq K} u_n e(nx) \right|^2 \leq C \sum_n |u_n|^2 (N + \delta^{-1}).$$

The problem remains to evaluate the constant C, at least asymptotically in N. (Elliott, 1985) gives a – next to trivial – proof of the inequality

$$\sum_{x \in \mathcal{X}} \max_{\substack{u < v \leq N, \\ v - u \leq H}} \left| \sum_{u \leq n \leq v} u_n e(nx) \right|^2 \leq \sum_n |u_n|^2 (H + 2\delta^{-1} \operatorname{Log}(e/\delta))$$

which is better in that the interval which the variable n ranges is arbitrarily located and further restrained in size. Furthermore, no implied constant appear, but the dependance in δ is worse. Montgomery's proof relies on Hunt's quantitative form of Carleson's theorem on almost sure convergence of L^2 Fourier series. As an effect, the constant C above is effective but no explicit version of it have been given – as of today, at least!

2 An extension of the classical arithmetical theory of the large sieve

Part of the material given here has already appeared in (Ramaré & Ruzsa, 2001). Theorem 2.1 is the main landmark of this chapter. From there onwards, what we do should become clearer to the reader. In particular, we shall detail an application of Theorem 2.1 to the Brun-Titchmarsh Theorem.

2.1. Sequences supported on compact sets

We introduce in this section some vocabulary that allows us handle modular arithmetic. All of it is trivial enough but will make life easier later on.

∘∘ By a *compact set* $\mathcal{K}$, we mean a sequence $\mathcal{K} = (\mathcal{K}_d)_{d \geq 1}$ satisfying

(1) $\mathcal{K}_d \subset \mathbb{Z}/d\mathbb{Z}$ for all $d \geq 1$.

(2) For any divisor d of q, we have $\sigma_{q \to d}(\mathcal{K}_q) = \mathcal{K}_d$ where $\sigma_{q \to d}$ is the canonical surjection (also called the restriction map) from $\mathbb{Z}/q\mathbb{Z}$ to $\mathbb{Z}/d\mathbb{Z}$:

$$(2.1) \qquad \sigma_{q \to d} : \quad \begin{array}{l} \mathbb{Z}/q\mathbb{Z} \to \mathbb{Z}/d\mathbb{Z} \\ x \bmod q \mapsto x \bmod d. \end{array}$$

When $\mathcal{K}$ is not empty, we have $\mathcal{K}_1 = \mathbb{Z}/\mathbb{Z}$. As examples, we can take $\mathcal{K}_d = \mathbb{Z}/d\mathbb{Z}$ for all d or $\mathcal{K}_d = \mathcal{U}_d$, where $\mathcal{U}_d$ is the set of invertible classes modulo d. The intersection and union of compact sets is again a compact set.

We can also consider $\mathcal{K}$ a subset of $\hat{\mathbb{Z}} = \varprojlim \mathbb{Z}/d\mathbb{Z}$, in which case it is indeed a compact set. Furthermore we shall sometimes consider $\mathcal{K}_d$ as a subset of $\mathbb{Z}$: the set of relative integers whose reduction modulo d falls inside $\mathcal{K}_d$.

∘∘ We say that the compact set $\mathcal{K}$ is *multiplicatively split* if for any d_1 and d_2 coprime positive integers, the Chinese remainder map

$$(2.2) \qquad \mathbb{Z}/d_1 d_2 \mathbb{Z} \longrightarrow \mathbb{Z}/d_1 \mathbb{Z} \times \mathbb{Z}/d_2 \mathbb{Z}$$

sends $\mathcal{K}_{d_1 d_2}$ onto $\mathcal{K}_{d_1} \times \mathcal{K}_{d_2}$. In this case, the sets $\mathcal{K}_{p^\nu}$ for prime p and $\nu \geq 1$ determine $\mathcal{K}$ completely. Notice that when $\mathcal{K}$ is multiplicatively split:

$$(2.3) \qquad |\mathcal{K}_{[d,d']}||\mathcal{K}_{(d,d')}| = |\mathcal{K}_d||\mathcal{K}_{d'}|$$

for any d and d', where $[d, d']$ is the lcm and (d, d') the gcd of d and d'. Here $|\mathcal{A}|$ stands for the cardinality of a set $\mathcal{A}$.

oo A compact set is said to be *squarefree* if

$$\mathcal{K}_q = \sigma_{q \to d}^{-1}(\mathcal{K}_d)$$

whenever d divides q and has the same prime factors. For instance, $\mathcal{U}$ is squarefree since being prime to q or to its *squarefree kernel* is the same.

oo A particularly successful hypothesis on $\mathcal{K}$ was introduced by (Johnsen, 1971) in the context of polynomials over a finite field and used in the case of the integers by (Gallagher, 1974) (see also (Selberg, 1976)). It reads

$$\forall d | q, \ \forall a \in \mathcal{K}_d$$

(2.4)   the quantity $\displaystyle\sum_{\substack{n \equiv a[d] \\ n \in \mathcal{K}_q}} 1$ is independent of a.

Another way to present this quantity would be to say it is the cardinality of $\sigma_{p^\nu \to p^{\nu-1}}^{-1}(\{a\})$. Since the introduction of this condition in our context is due to (Gallagher, 1974), we shall refer to it as the Johnsen-Gallagher condition. Note that this condition does not require $\mathcal{K}$ to be multiplicatively split, although all our examples will also satisfy this additional hypothesis.

Any squarefree compact set automatically satisfies the Johnsen-Gallagher hypothesis. Since the sieve kept to such sets for a very long time, and the combinatorial sieve still does, this condition does not show up in classical expositions. We present in Theorem 13.1 a result that is unreachable if we were to confine ourselves to squarefree sieves.

2.2. A family of arithmetical functions

Let us start with a multiplicatively split compact set $\mathcal{K}$. We consider the non-negative multiplicative function h defined by

(2.5) $$h(d) = \prod_{p^\nu \| d} \left(\frac{p^\nu}{|\mathcal{K}_{p^\nu}|} - \frac{p^{\nu-1}}{|\mathcal{K}_{p^{\nu-1}}|} \right) \geq 0, \quad h(1) = 1$$

where $q \| d$ means that q divides d in such a way that q and d/q are coprime. We shall say that q *divides d exactly*. Note that

(2.6) $$\frac{d}{|\mathcal{K}_d|} = \sum_{\delta | d} h(\delta).$$

We further define

(2.7)
$$G_d(Q) = \sum_{\substack{\delta \leq Q, \\ [d,\delta] \leq Q}} h(\delta)$$

which we also denote by $G_d(\mathcal{K}, Q)$ when mentioning the compact set $\mathcal{K}$ is of any help. Let us note that in the extremal case $\mathcal{K}_d = \mathbb{Z}/d\mathbb{Z}$, we have $h(d) = 0$ except when $d = 1$ in which case we have $h(1) = 1$. This implies that $G_d(Q) = 1$ for all d's. These fairly unusual functions appear in the following form:

Lemma 2.1. *We have*

$$G_d(Q) = \sum_{\substack{q \leq Q \\ d|q}} \left(\sum_{f/d|f|q} \mu(q/f) f/|\mathcal{K}_f| \right).$$

This is easily proved using (2.6). We present in chapter 3 a more abstracted approach to this set of functions.

Often, the set $\mathcal{K}$ is squarefree, in which case the above expression simplifies and we recognize, up to a factor, the usual functions from the Selberg sieve (see (2.8) below). In particular, we know how to evaluate them. We shall give two examples of such an evaluation in sections 2.4 and 5.4 and a general theorem in the Appendix. The reader should consult (Levin & Fainleib, 1967), (Halberstam & Richert, 1971) and (Halberstam & Richert, 1974) for the general theory. Meanwhile, we move to another lemma.

Lemma 2.2. *We have*

$$|\mathcal{K}_d| \sum_{\substack{q \leq Q \\ d|q}} \mu(q/d) G_q(Q) = d \sum_{\substack{\ell \leq Q \\ d|\ell}} \mu(\ell/d).$$

We refer to section 11.3 for an interpretation of the above lemma and background information on how it came to be.

Proof. We appeal to Lemma 2.1 and write:

$$\sum_{\substack{q \leq Q \\ d|q}} \mu(q/d) G_q(Q) = \sum_{\substack{q \leq Q \\ d|q}} \mu(q/d) \sum_{\substack{\ell \leq Q \\ q|\ell}} \left(\sum_{q|f|\ell} \mu(\ell/f) f/|\mathcal{K}_f| \right)$$

$$= \sum_{\substack{\ell \leq Q \\ d|\ell}} \sum_{d|f|\ell} \mu(\ell/f) \frac{f}{|\mathcal{K}_f|} \sum_{\substack{q \leq Q \\ d|q|f}} \mu(q/d)$$

in which only the term $d = f$ remains, thus proving our assertion. $\diamond\diamond\diamond$

We conclude by a lemma that is in fact a generalization of a lemma of (van Lint & Richert, 1965) but which is trivial in our setting.

Lemma 2.3. *We have $G_\ell(Q\ell/d) \le G_d(Q) \le G_\ell(Q)$ for $\ell | d$.*

When the compact set is squarefree, the reader will check from (2.5) that $h(d) = 0$ as soon as d is not squarefee. In that case, the summand appearing in Lemma 2.1 vanishes whenever q/d and d are coprime. We can thus write $q = d\ell$ with $(\ell, d) = 1$ in this Lemma, which leads to (see also (5.9))

$$(2.8) \qquad G_d(Q) = \frac{d}{|\mathcal{K}_d|} \sum_{\substack{\ell \le Q/d \\ (\ell,d)=1}} h(\ell).$$

Since in classical literature $\mathcal{K}$ is always squarefree, authors tend to call $G_d(Q)$ what is in fact $|\mathcal{K}_d| G_d(Q)/d$ in our notation. We had the option of introducing another name, but we prefered to retain the same name in these lectures, for the reason that the most important value $G_1(Q)$ is unchanged. Note that it is usual to simply denote this latter value by $G(Q)$, a usage that we avoid.

2.3. An identity

We say that the sequence $(u_n)_{n \ge 1}$ of complex numbers is *carried by* $\mathcal{K}$ *up to level* Q when the support of $(u_n)_{n \ge 1}$ belongs to $\mathcal{K}_q$ for all $q \le Q$, or formally:

$$(2.9) \qquad u_n \ne 0 \implies \forall q \le Q, n \in \mathcal{K}_q.$$

As examples, note that every sequence is carried by $(\mathbb{Z}/q\mathbb{Z})_{q \ge 1}$ up to any level, and that the sequence of primes $> Q$ is carried by $\mathcal{U}$ up to level Q.

Here is a generalization of known identities, see (Rényi, 1958), (Rényi, 1959), (Bombieri & Davenport, 1968), (Montgomery, 1971) as well as (Bombieri, 1987):

Theorem 2.1. *When $\mathcal{K}$ is multiplicatively split and verifies the Johnsen-Gallagher condition (2.4) and (u_n) is a sequence carried by $\mathcal{K}$ up to level Q we have*

$$\sum_{q \le Q} G_q(Q) |\mathcal{K}_q| \sum_{b \in \mathcal{K}_q} \left| \sum_{\ell | q} \mu\left(\frac{q}{\ell}\right) \frac{|\mathcal{K}_\ell|}{|\mathcal{K}_q|} \sum_{m \equiv b[\ell]} u_m \right|^2 = \sum_{q \le Q} \sum_{a \bmod^* q} \left| \sum_n u_n e\left(\frac{na}{q}\right) \right|^2.$$

The same identity holds true but with the set $\{q \leq Q\}$ replaced by any set $\mathcal{Q}$ of moduli closed under division, by which we mean that if $q \in \mathcal{Q}$ and $d|q$ then d is also in $\mathcal{Q}$. It is easy to see, simply by following the proof below, that condition (2.4) is indeed required. Note that in order to handle the non-square-free q, a proper definition of G_q is needed.

Proof. Let $\Delta(Q)$ be the R.H.S. of the above equality. We have

$$\Delta(Q) = \sum_{m,n} u_m \overline{u_n} \sum_{d|m-n} d \sum_{q \leq Q/d} \mu(q).$$

On using Lemma 2.2 to modify the inner sum we obtain

$$\Delta(Q) = \sum_q G_q(Q) \left\{ \sum_{d|q} \mu(q/d)|\mathcal{K}_d| \sum_{m \equiv n[d]} u_m \overline{u_n} \right\}.$$

Let us set

$$(2.10) \qquad \Theta(q) = |\mathcal{K}_q| \sum_{b \in \mathcal{K}_q} \left| \sum_{\ell|q} \mu(q/\ell) \frac{|\mathcal{K}_\ell|}{|\mathcal{K}_q|} \sum_{m \equiv b[\ell]} u_m \right|^2.$$

On expanding the square we get

$$\Theta(q) = \sum_{m,n} u_m \overline{u_n} \sum_{\ell_1|q, \ell_2|q} \mu(q/\ell_1) \mu(q/\ell_2) \frac{|\mathcal{K}_{\ell_1}||\mathcal{K}_{\ell_2}|}{|\mathcal{K}_q|} \sum_{\substack{b \in \mathcal{K}_q, \\ m \equiv b[\ell_1], \\ n \equiv b[\ell_2]}} 1.$$

We introduce $d = (\ell_1, \ell_2)$. Our conditions imply that $m \equiv n[d]$. Once this is guaranted, b is determined modulo $[\ell_1, \ell_2]$ by m and n; the Johnsen-Gallagher condition (2.4) then implies that there are $|\mathcal{K}_q|/|\mathcal{K}_{[\ell_1,\ell_2]}|$ choices for b. Recalling (2.3), we reach

$$\Theta(q) = \sum_{d|q} \sum_{m \equiv n[d]} u_m \overline{u_n} |\mathcal{K}_d| \sum_{\substack{\ell_1|q, \ell_2|q \\ (\ell_1, \ell_2)=d}} \mu(q/\ell_1) \mu(q/\ell_2).$$

We are left with computing the most inner sum which is readily done:

$$\sum_{\substack{\ell_1|q, \ell_2|q \\ (\ell_1,\ell_2)=d}} \mu(q/\ell_1) \mu(q/\ell_2) = \sum_{\substack{r_1|q/d \\ r_2|q/d}} \mu((q/d)/r_1) \mu((q/d)/r_2) \sum_{\substack{\delta|r_1 \\ \delta|r_2}} \mu(\delta)$$

$$= \mu(q/d)$$

as required.   ◇ ◇ ◇

To understand the L.H.S. of this theorem, consider the case $\mathcal{K}_d = \mathbb{Z}/d\mathbb{Z}$ due to (Montgomery, 1968) but reduce it to the case when $q = p$ a prime numberas in (Rényi, 1958). We get
(2.11)

$$|\mathcal{K}_p| \sum_{b \in \mathcal{K}_p} \left| \sum_{\ell \mid p} \mu(p/\ell) \frac{|\mathcal{K}_\ell|}{|\mathcal{K}_p|} \sum_{m \equiv b[\ell]} u_m \right|^2 = p \sum_{b \bmod p} \left| \sum_{m \equiv b[p]} u_m - \frac{\sum_m u_m}{p} \right|^2$$

so this quantity measures the distortion from equidistribution in arithmetic progressions. This is also true of the quantity with general q, as the reader will realize after some thought. However, if we know the sequence can only reach some congruence classes, namely the ones in some $\mathcal{K}_p$, then the proper approximation is $\sum_m u_m/|\mathcal{K}_p|$ and not $\sum_m u_m/p$. This is what is put in place in the above result. In chapter 4 we provide a more geometrical interpretation.

We recover in this manner a theorem of (Gallagher, 1974). This is an analogue of a similar theorem proved in (Johnsen, 1971) in the context of polynomials over finite fields.

Corollary 2.1 (Gallagher). *Assume $\mathcal{K}$ is multiplicatively split and verifies the Johnsen-Gallagher condition (2.4). Let Z denotes the number of integers in the interval $[M + 1, M + N]$ that belongs to $\mathcal{K}_d$ for all $d \leq Q$. We have*

$$Z \leq (N + Q^2)/G_1(Q).$$

It was (Bombieri & Davenport, 1968) who first used the large sieve to get this kind of result, namely for primes, and (Montgomery, 1968) worked out a general theorem along lines closer to that of (Rényi, 1958). We derive some classical bounds from this inequality in the Appendix. It will also give the reader the opportunity to manipulate the concept of a compact set in connection with sieve problems.

Proof. We take for $(u_n)_{n \geq 1}$ the characteristic function of the set whose cardinality is to be evaluated and apply Theorem 2.1 together with the large sieve inequality. We finally discard all terms on the L.H.S. except the one corresponding to $q = 1$.   ◇◇◇

Note that (Selberg, 1976) proves a similar theorem but without the Johnsen-Gallagher condition. We shall do so in chapter 13, this time enabling also the sieving of a general sequence instead of an interval, but note that our present way of doing offers what is sometimes known as *a large sieve extension* of this bound, in the spirit of the theorem of (Bombieri & Davenport, 1968) we recall in section 2.5. See also Theorem 15.3 for a generalization in another direction.

2.4. The Brun-Titchmarsh Theorem

This theorem reads as follows:

Theorem 2.2. *Let $M \geq 0$ and $N > q \geq 1$ be given and let a be an invertible residue class modulo q. The number Z of primes in the interval $[M + 1, M + N]$ lying in the residue class a modulo q verifies*

$$Z \leq \frac{2N}{\phi(q)\operatorname{Log}(N/q)}.$$

This neat and effective version is due to (Montgomery & Vaughan, 1973). Earlier versions essentially had $2 + o(1)$ instead of simply 2. The name "Brun-Titchmarsh" Theorem stems from (Linnik, 1961). Indeed, Titchmarsh proved such a theorem for $q = 1$ with a $\operatorname{Log}\operatorname{Log}(N/q)$ term instead of the 2 to establish the asymptotic for the number of divisors of the $p + 1$, p ranging through the primes, and he used the method of Brun. The constant 2 (with a $o(1)$) appeared for the first time in (Selberg, 1949).

To clarify the argument we restrict our attention to the case $q = 1$ and get $2 + o(1)$ instead of 2. Start with Corollary 2.1 applied to $\mathcal{K} = \mathcal{U}$. To make this possible we restrict our attention to primes $> Q$. We then find that

(2.12) $|\mathcal{K}_d| = \phi(d)$, and $h(d) = \mu^2(d)/\phi(d)$.

So we are left with finding a lower bound for $G_1(Q)$. Write

$$\frac{\mu^2(d)}{\phi(d)} = \frac{\mu^2(d)}{d} \prod_{p|d} \frac{1}{1 - \frac{1}{p}} = \frac{\mu^2(d)}{d} \prod_{p|d} \left(1 + \frac{1}{p} + \frac{1}{p^2} + \dots\right)$$

$$= \mu^2(d) \sum_{\substack{k \geq 1, d|k \\ [p|k \implies p|d]}} \frac{1}{k}$$

which we sum to get

(2.13) $G_1(Q) = \displaystyle\sum_{d \leq Q} \mu^2(d) \sum_{\substack{k \geq 1, d|k \\ [p|k \implies p|d]}} \frac{1}{k} \geq \sum_{k \leq Q} \frac{1}{k} \geq \operatorname{Log} Q.$

It can be fairly easily shown that in fact $G_1(Q) = \operatorname{Log} Q + \mathcal{O}(1)$, either by reading section 5.3 or by applying Theorem 21.1 from the appendix.

We now choose $Q = \sqrt{N}/\operatorname{Log} N$, getting

(2.14) $Z \leq \dfrac{2N(1 + \mathcal{O}(\operatorname{Log}^{-2} N))}{\operatorname{Log} N - 2\operatorname{Log}\operatorname{Log} N} + Q$

which is indeed not more than $2(1+o(1))N/\operatorname{Log} N$. To prove the theorem for primes in a residue class, sieve the arithmetic progression $a+mq$, where m varies in an interval, up to a level $Q = \sqrt{N/q}/\operatorname{Log}(N/q)$.

2.5. The Bombieri-Davenport Theorem

This section is somewhat astray from our main line but deserves a place since it is this result that led the author to believe that something like Theorem 2.1 ought to exist.

Theorem 2.3 (Bombieri & Davenport). *When $(u_n)_{n \leq N}$ is such that u_n vanishes as soon as n has a prime factor less than Q, we have*

$$\sum_{q \leq Q} \operatorname{Log}(Q/q) \sum_{\chi \bmod^* q} \left| \sum_n u_n \chi(n) \right|^2 \leq \sum_n |u_n|^2 (N + Q^2)$$

where $\chi \bmod^* q$ denotes a summation over all primitive characters modulo q.

With $\mathcal{K} = \mathcal{U}$ and our terminology above, the hypothesis says that (u_n) is carried by $\mathcal{K}$ upto the level Q. We now deduce this result from Theorem 2.1.

Proof. We first show that what we termed $\Theta(q)$ in (2.10) is in fact the summand of the L.H.S. above. When χ is a character, we denote its conductor by f_χ. On detecting the congruence condition $m \equiv b[\ell]$ using multiplicative characters (this is possible because b and u_n are prime to ℓ), we get for any fixed multiple q of ℓ:

$$\sum_{m \equiv b[\ell]} u_m = \frac{1}{\phi(\ell)} \sum_{\chi \bmod \ell} \sum_m \overline{\chi(b)} \chi(m) u_m$$

$$= \frac{1}{\phi(\ell)} \sum_{\substack{\chi \bmod q \\ f_\chi | \ell}} \sum_m \overline{\chi(b)} \chi(m) u_m.$$

From which we easily deduce

$$\sum_{\ell/\ell | q} \mu(q/\ell) \frac{\phi(\ell)}{\phi(q)} \sum_{m \equiv b[\ell]} u_m = \sum_{\ell | q} \frac{\mu(q/\ell)}{\phi(q)} \sum_{\substack{\chi \bmod q \\ f_\chi | \ell}} \sum_m \overline{\chi(b)} \chi(m) u_m$$

$$= \frac{1}{\phi(q)} \sum_{\chi \bmod^* q} \sum_m \overline{\chi(b)} \chi(m) u_m.$$

Squaring this quantity, summing it over all reduced residue classes modulo q and multiplying the result by $\phi(q)$ indeed gives

$$\Theta(q) = \sum_{\chi \bmod^* q} \left| \sum_m \chi(m) u_m \right|^2 .$$

This last step amounts to applying Plancherel formula on $(\mathbb{Z}/d\mathbb{Z})^*$. To find a lower bound for the factor $G_q(Q)$ we use Lemma 2.3 and get $G_q(Q) \geq G_1(Q/q)$, which using (2.13) this is indeed $\geq \mathrm{Log}(Q/q)$. The theorem now follows.   ◇◇◇

The proof that Bombieri & Davenport gave uses the value of the Gauss sums, and my first motivation was to remove this part, since it seemed clear, it was only a matter of orthonormal systems. Then the multiplicativity of these characters is not used either and back in 1992, I started developing a general theory of "characters" to prove a similar result. This was however not very convenient because I had to explain what these were; after having understood the Selberg sieve in a similar setting, something we shall do in chapter 11, I finally found the identity of Theorem 2.1 with a proof from which my abstract characters disappeared.

Note further that it is not enough to substitute Theorem 1.2 to Theorem 1.1 to get a *maximal* variant of this theorem (i.e. a result in which the $|\sum_n u_n \chi(n)|$ would be replaced by $\max_{K \leq N} |\sum_{1 \leq n \leq K} u_n \chi(n)|$). See (Elliott, 1991).

The strength of this theorem seems to have been underestimated, and we conclude on this aspect, somewhat anticipating the proof of Theorem 5.2. (Elliott, 1983) improving on (Elliott, 1977) proves that

$$\sum_{\substack{q \leq Q, \\ q \text{ prime}}} (q-1) \sum_{a \bmod^* q} \left| \sum_{\substack{p \leq N, \\ p \equiv a[q]}} u_p - \frac{\sum_{p \leq N} u_p}{q-1} \right|^2$$

$$\ll \left(\frac{N}{\mathrm{Log}\, N} + Q^{54/11+\varepsilon} \right) \sum_{p \leq N} |u_p|^2 .$$

As it turns out, the summand is simply $\sum_{\chi \bmod^* q} |\sum_{p \leq N} u_p \chi(p)|^2$, the only non primitive character being the principal one, since q is prime. We can thus use the Bombieri-Davenport Theorem up to level $\sqrt{N}$ and restrict then summation to $q \leq Q$ (as in the proof of Theorem 5.2 below), getting the upper bound

$$\frac{2N}{\mathrm{Log}(\sqrt{N}/Q)} \sum_{p \leq N} |u_p|^2$$

instead of the above, which allows Q up to $N^{1/2-\varepsilon}$. Note further that in this approach, we may replace the set $p \leq N$, by any set of primes in an interval of length N.

Theorem 2 and Corollary 4 of (Puchta, 2003)[1] follow similarly from this same remark, since this author directly discusses primitive character sums. However, the methods used therein apply also to shorter sets of characters modulo a single modulus, and are now beyond the present approach. They still belong to the realm of almost orthogonality, and Lemma 1.1 is still being used, but with fine character sum bounds.

2.6. A detour towards lower bounds

The L.H.S. of Theorem 2.1 will be very small when our sequence is very well distributed in arithmetic progressions. On an other hand, the R.H.S. may be expected to approximate $\sum_n |u_n|^2 Q^2$, if one follows for instance the proof in terms of Riemann sums given by (Gallagher, 1967). Indeed (Roth, 1964) proved that dense sequences that are not too dense could not be evenly distributed in arithmetic progressions. (Huxley, 1972b) strengthened this work to the case of neither too thin nor too dense *sifted sequences*, by which we mean a sequence whose characteristic function is "carried" – see (2.9)– by some squarefree compact set. The proof goes by finding a lower bound for a certain variance expression. It seems plausible that with ideas from the proof of Theorem 2.1, one can extend this result to the case of non-squarefree compact sets verifying the Johnsen-Gallagher condition, and that one could also introduce a more precise kind of "variance" expression. See also section 20.7 for a reversed large sieve inequality.

---

[1] I had very interesting discussions with J.-C. Puchta in spring 2006 on this very subject, which is how I got to notice what I call here an "underestimation".

3 Some general remarks on arithmetical functions

We present here some general material pertaining to the family of functions we consider in our sieve setting (see chapter 2, in particular section 2.2).

When $d \geq 1$ is an integer, let us write δ_d to denote the arithmetical function which takes the value 1 at d and the value 0 at all other integers ≥ 1. Let $\mathbb{1}_{d \cdot \mathbb{N}}$ denote the arithmetical function $1 \star \delta_d$. It is easily verified that $\mathbb{1}_{d \cdot \mathbb{N}}$ is the characteristic function of the set of multiples of d and that $(\mu \star \delta_d)(m) = \mu(m/d)\mathbb{1}_{d \cdot \mathbb{N}}(m)$, for all $m \geq 1$.

We recall that a subset X of the integers ≥ 1 is said to closed under division if every divisor of each element of X is also in X. We write $\mathscr{A}(X)$ to denote the set of complex valued functions on X. It is easily seen that $\mathscr{A}(X)$ is a commutative ring with respect to addition and (dirichlet) convolution.

Lemma 3.1. *Let X be a subset of the integers ≥ 1 that is closed under division. Let ϕ be in $\mathscr{A}(X)$ and let $\psi = \mu \star \phi$. For all f and g finitely supported functions in $\mathscr{A}(X)$ we have the identities*

$$(3.1) \quad \sum_{k \in X} f(k)\phi(k) = \sum_{m \in X} \psi(m) \sum_{\substack{k \in X, \\ m|k}} f(k) = \sum_{m \in X} \sum_{\substack{k \in X, \\ m|k}} \psi(k/m)f(k)$$

and

$$(3.2) \quad \sum_{k \in X} \sum_{\ell \in X} f(k)g(\ell)\phi((k,\ell)) = \sum_{m \in X} \psi(m) \sum_{\substack{k \in X, \\ m|k}} f(k) \sum_{\substack{\ell \in X, \\ m|\ell}} g(\ell).$$

Equation (3.2) is the heart of the Selberg diagonalization process, as it is used for instance in section 11.3.

Proof. Since f and g are finitely supported and since all terms in (3.1) are linear in f and both sides in (3.2) are bilinear in f and g, it suffices to verify these relations when $f = \delta_a$ and $g = \delta_b$, for any integers $a, b \in X$. When this is the case, and since X is divisor closed, these relations reduce respectively to the obvious relations

$$\phi(a) = \sum_{m|a} \psi(m) = \sum_{m|a} \psi(a/m) \quad \text{and} \quad \phi((a,b)) = \sum_{\substack{m, \\ m|a, m|b}} \psi(m).$$

◇ ◇ ◇

Corollary 3.1. *Let a be an integer ≥ 1 and d a divisor of a. We then have that*

$$(3.3) \qquad \delta_d(a) = \sum_{\substack{k|a,\\ d|k}} \mu(a/k) = \sum_{\substack{k|a,\\ d|k}} \mu(k/d).$$

Proof. We apply (3.1) with $f = \delta_a$ and $\phi = \delta_d$ and X the set of divisors of a.   ◇◇◇

Corollary 3.2. *Let X be a subset of the integers ≥ 1 that is closed under division and d be an integer in X. For any finitely supported function f in $\mathscr{A}(X)$ we have*

$$(3.4) \qquad f(d) = \sum_{\substack{k \in X, \\ d|k}} \sum_{\substack{q \in X, \\ k|q}} \mu(q/k)f(q) = \sum_{\substack{k \in X, \\ d|k}} \mu(k/d) \sum_{\substack{q \in X, \\ k|q}} f(q).$$

Proof. It suffices to verify (3.4) when f is of the form δ_a, for any integer $a \in X$. When this is the case, and because X is closed under division and $\delta_a(d) = \delta_d(a)$, (3.4) reduces to (3.3).   ◇◇◇

Corollary 3.3. *Let q be an integer ≥ 1 and d be a divisor of q. We then have the relation*

$$(3.5) \qquad \sum_{\substack{k|q, \ell|q, \\ (k,\ell)=d}} \mu(q/k)\mu(q/\ell) = \mu(q/d).$$

Proof. We apply (3.2) with X taken to be the set of divisors of q, f and g both taken to be the function $k \mapsto \mu(q/k)$ on X and $\phi = \delta_d$. Then $\psi(m) = \mu(m/d)\mathbb{1}_{d \cdot \mathbb{N}}(m)$ and, using (3.3), the right hand side of (3.2) reduces to $\mu(m/d)\mathbb{1}_{d \cdot \mathbb{N}}(m)\delta_m(q) = \mu(q/d)$.   ◇◇◇

Let f be an arithmetical function and Q be a real number ≥ 1. for each integer d in the interval $[1, Q]$ we define

$$(3.6) \qquad G_d(f, Q) = \sum_{q/[d,q] \leq Q} f(q).$$

This set of functions will be required to define the λ_d's of section 13.1.

Corollary 3.4. *Let f be an arithmetical function and let $g = \mathbb{1} \star f$. When Q is a real number ≥ 1, and for each integer d in $[1, Q]$, we have*

$$(3.7) \qquad G_d(f, Q) = \sum_{\substack{k \geq 1, \\ d|k}} g(k) \sum_{\substack{q \leq Q, \\ k|q \leq Q}} \mu(q/k).$$

Proof. We reduce to the case when $f = \delta_a$, where a is an integer ≥ 1. Then $g = \mathbb{1}_{a \cdot \mathbb{N}}$. On writing χ_Q to denote the characteristic function of the integers in the interval $[1, Q]$ and using (3.4) with X taken to be the set of all integers ≥ 1 we then have

$$\sum_{[q,d] \leq Q} \delta_a(q) = \chi_Q([a, d])$$

$$= \sum_{\substack{k \geq 1, \\ [a,d] \mid k}} \sum_{\substack{q, \\ k \mid q}} \mu(q/k) \chi_Q(q) = \sum_{\substack{k \geq 1, \\ d \mid k}} \mathbb{1}_{a \cdot \mathbb{N}}(k) \sum_{\substack{q \leq Q, \\ k \mid q}} \mu(q/k).$$

$\diamond \diamond \diamond$

Corollary 3.5. *Let f be an arithmetical function and let $g = \mathbb{1} \star f$. When Q is a real number ≥ 1, and d is an integer in $[1, Q]$, we have*

$$\sum_{\substack{q \geq 1, \\ d \mid q}} \mu(q/d) G_q(f, Q) = g(d) \sum_{\substack{q \leq Q, \\ d \mid q}} \mu(q/d).$$

Proof. Since the arithmetical function

$$k \longmapsto g(k) \sum_{\substack{q \leq Q, \\ k \mid q}} \mu(q/k)$$

vanishes when $k > Q$, it is of finite support. Thus the corollary follows from (3.7) and (3.4) applied with X taken to be the set of all integers ≥ 1.

$\diamond \diamond \diamond$

4 A geometrical interpretation

The expression appearing in Theorem 2.1 may look unpalatable, but is in fact simply the norm of a suitable orthonormal projection, as we show here. The reader may skip this chapter. While it does different insights on what we are doing, it will not be invoked before chapter 19, with two short detours at sections 9.4 and 11.4.

Throughout this chapter, we fix a **multiplicatively split** compact set $\mathcal{K}$ verifying the Johnsen-Gallagher condition (2.4). For fixed q let $\mathcal{F}(\mathcal{K}_q)$ be the vector space of complex valued functions over $\mathcal{K}_q$. Such functions may also be seen as functions over $\mathbb{Z}/q\mathbb{Z}$ that vanish outside $\mathcal{K}_q$. We endow this vector space with a hermitian product by setting

$$(4.1) \qquad [f|g]_q = \frac{1}{|\mathcal{K}_q|} \sum_{n \bmod q} f(n)\overline{g(n)}.$$

We should emphasize that the split multiplicativity is an essential part of the present study. In terms of sieving as the problem is exposed in chapter 11 the compact set $\mathcal{K}$ corresponds to the host sequence and thus will often be taken to be $\hat{\mathbb{Z}}$ also denoted by $(\mathbb{Z}/d\mathbb{Z})_d$ depending on the definition you prefer. But we have seen in Bombieri & Davenport's approach how the host sequence could become the sifted one (see sections 2.3 and 2.5)!

4.1. Local couplings

Our first task is to link together the arithmetic modulo distinct moduli. To do so, we consider the usual lift when $d|q$:

$$L_{\tilde{q}}^{\tilde{d}} : \mathcal{F}(\mathcal{K}_d) \to \mathcal{F}(\mathcal{K}_q)$$

$$(4.2) \qquad f \mapsto f \circ \sigma_{q \to d} : \mathcal{K}_q \to \mathbb{C}$$
$$x \mapsto f(x \bmod d)$$

This function is a natural one. The reader may wonder why we chose $\tilde{q}$ instead of q; it will avoid troubles latter on. In order to further compare the hermitian structures, we consider the operator $J_{\tilde{d}}^{\tilde{q}}$ from $\mathcal{F}(\mathcal{K}_q)$ to $\mathcal{F}(\mathcal{K}_d)$ which associates to $f \in \mathcal{F}(\mathcal{K}_q)$ the function

$$(4.3) \qquad J_{\tilde{d}}^{\tilde{q}}(f) : \mathcal{K}_d \to \mathbb{C}, \qquad x \mapsto \frac{|\mathcal{K}_d|}{|\mathcal{K}_q|} \sum_{\substack{n \in \mathcal{K}_q, \\ n \equiv x[d]}} f(n).$$

This operator verifies the fundamental:

$$(4.4) \qquad [L_{\tilde{q}}^{\tilde{d}}(f)|g]_q = [f|J_{\tilde{d}}^{\tilde{q}}(g)]_d.$$

Proof. We simply check directly that

$$[L_{\tilde{q}}^{\tilde{d}}(f), g]_q = \frac{1}{|\mathcal{K}_q|} \sum_{x \in \mathcal{K}_d} f(x) \sum_{\substack{n \in \mathcal{K}_q, \\ n \equiv x[d]}} \overline{g(n)}$$

$$= \frac{1}{|\mathcal{K}_d|} \sum_{x \in \mathcal{K}_d} f(x) \overline{\left(\frac{|\mathcal{K}_d|}{|\mathcal{K}_q|} \sum_{\substack{n \in \mathcal{K}_q, \\ n \equiv x[d]}} g(n) \right)}$$

as required.   ◇◇◇

Thus the maps $L_{\tilde{q}}^{\tilde{d}}$ and $J_{\tilde{d}}^{\tilde{q}}$ are adjoint one to another, even if the reader may be unfamiliar with the concept when applied to linear functions that are *not* homomorphisms! Let us define

$$(4.5) \qquad U_{\tilde{q} \to \tilde{d}} = L_{\tilde{q}}^{\tilde{d}} J_{\tilde{d}}^{\tilde{q}}.$$

The next section is devoted to understanding these operators. Note that they depend on $\mathcal{K}$ even if our notation does not make this apparent.

4.2. The Fourier structure

We start with the following fundamental property.

Lemma 4.1. *The operator $U_{\tilde{q} \to \tilde{d}}$ is hermitian. Furthermore, $U_{\tilde{q} \to \tilde{d}_1}$ and $U_{\tilde{q} \to \tilde{d}_2}$ commute with each other and we have*

$$(4.6) \qquad U_{\tilde{q} \to \tilde{d}_1} U_{\tilde{q} \to \tilde{d}_2} = U_{\tilde{q} \to \widetilde{(d_1, d_2)}}.$$

Proof. The hermitian character is readily proved:

$$[U_{\tilde{q} \to \tilde{d}}(f)|g]_q = [L_{\tilde{q}}^{\tilde{d}} J_{\tilde{d}}^{\tilde{q}}(f)|g]_q = [J_{\tilde{d}}^{\tilde{q}}(f)|J_{\tilde{d}}^{\tilde{q}}(g)]_q$$
$$= \overline{[J_{\tilde{d}}^{\tilde{q}}(g)|J_{\tilde{d}}^{\tilde{q}}(f)]_q} = \overline{[L_{\tilde{q}}^{\tilde{d}} J_{\tilde{d}}^{\tilde{q}}(g)|f]_q} = [f|L_{\tilde{q}}^{\tilde{d}} J_{\tilde{d}}^{\tilde{q}}(g)]_q,$$

where, in fact, we have not used any property of $\mathcal{K}$. The commuting property requires more hypothesis. By using the definition of $U_{\tilde{q} \to \tilde{d}_1}$, we find that

$$U_{\tilde{q} \to \tilde{d}_1} U_{\tilde{q} \to \tilde{d}_2}(f)(x) = \frac{|\mathcal{K}_{d_1}|}{|\mathcal{K}_q|} \sum_{\substack{n \in \mathcal{K}_q, \\ n \equiv x[d_1]}} U_{\tilde{q} \to \tilde{d}_2}(f)(n)$$

into which we plug the definition of $U_{\tilde{q}\to d_2}$ to reach

$$U_{\tilde{q}\to d_1} U_{\tilde{q}\to d_2}(f)(x) = \frac{|\mathcal{K}_{d_1}|}{|\mathcal{K}_q|} \sum_{\substack{n\in\mathcal{K}_q,\\ n\equiv x[d_1]}} \frac{|\mathcal{K}_{d_2}|}{|\mathcal{K}_q|} \sum_{\substack{m\in\mathcal{K}_q,\\ m\equiv n[d_2]}} f(m)$$

$$= \frac{|\mathcal{K}_{d_1}||\mathcal{K}_{d_2}|}{|\mathcal{K}_q|} \sum_{m\in\mathcal{K}_q} W(m;x)f(m),$$

say, where we have written $W(m;x)$ to denote

(4.7)
$$W(m;x) = \sum_{\substack{n\in\mathcal{K}_q,\\ n\equiv x[d_1], n\equiv m[d_2]}} 1.$$

The reader will check that $W(m;x) = 0$ when m and x are not congruent modulo (d_1, d_2) and that $W(m;x) = |\mathcal{K}_{d_1}||\mathcal{K}_{d_2}|/|\mathcal{K}_{(d_1,d_2)}|$ when they are. This proves (4.6), and hence the fact that the operators $U_{\tilde{q}\to \tilde{d}}$ commute with each other. Note that this argument depends crucially on the split multiplicativity of $\mathcal{K}$.   ◇◇◇

A consequence of the above lemma is that $U_{\tilde{q}\to \tilde{d}}$ is a hermitian projection. Let us further define

(4.8)
$$U_{\tilde{q}\to d} = \sum_{\ell|d} \mu(d/\ell) U_{\tilde{q}\to \tilde{\ell}}.$$

The main structure Theorem is the following:

Theorem 4.1. *The operators $(U_{\tilde{q}\to d})_{d|q}$ are two by two orthogonal hermitian projections. For each divisor r of q, we further have*

$$U_{\tilde{q}\to \tilde{r}} = \sum_{d|r} U_{\tilde{q}\to d}.$$

Note that $U_{\tilde{q}\to \tilde{q}}$ is the identity.

Proof. On applying the preceding lemma, we get

$$U_{\tilde{q}\to d_1} U_{\tilde{q}\to d_2} = \sum_{\ell_1|d_1, \ell_2|d_2} \mu(d_1/\ell_1)\mu(d_2/\ell_2) U_{\tilde{q}\to \widetilde{(\ell_1,\ell_2)}}$$

$$= \sum_{t|(d_1,d_2)} \left(\sum_{\substack{\ell_1|d_1,\ell_2|d_2\\ (\ell_1,\ell_2)=t}} \mu(d/\ell_1)\mu(d/\ell_2) \right) U_{\tilde{q}\to \tilde{t}}.$$

The inner coefficient is multiplicative and is readily seen to vanish when $d_1 \neq d_2$ and to equal $\mu(d_1/t)$ otherwise, thus establishing that $U_{\tilde{q}\to d}$ is indeed a projection and that $U_{\tilde{q} \to d_1}$ and $U_{\tilde{q} \to d_2}$ are orthonormal when $d_1 \neq d_2$. The remaining statements follow. $\diamond\diamond\diamond$

Theorem 4.1 is the main basis of what now follows. Let us set

(4.9) $$\mathfrak{M}(\tilde{q} \to d) = U_{\tilde{q} \to d} \mathscr{F}(\mathcal{K}_q)$$

which we endow with the scalar product of $\mathscr{F}(\mathcal{K}_q)$. This set depends on q: it is made of functions over $\mathcal{K}_q$ but this dependence is immaterial since

Lemma 4.2.

(4.10) $$\mathfrak{M}(\tilde{q} \to d) \underset{L_{\tilde{q}}^{\tilde{d}}}{\overset{J_{\tilde{d}}^{\tilde{q}}}{\rightleftarrows}} \mathfrak{M}(\tilde{d} \to d)$$

are isometries, inverses of each other.

This lemma legitimates a special name for $\mathfrak{M}(\tilde{d} \to d)$, which we simply call $\mathfrak{M}(d)$.

Proof. We first note that $L_{\tilde{q}}^{\tilde{d}} U_{\tilde{d} \to d} J_{\tilde{d}}^{\tilde{q}}(F) = U_{\tilde{d} \to d}(F)$, which in passing proves that $L_{\tilde{q}}^{\tilde{d}} \mathfrak{M}(\tilde{d} \to d) = \mathfrak{M}(\tilde{q} \to d)$. Next, given any two elements $U_{\tilde{d} \to d}(f)$ and $U_{\tilde{d} \to d}(g)$ of $\mathfrak{M}(\tilde{d} \to d)$, we have

$$[L_{\tilde{q}}^{\tilde{d}} U_{\tilde{d} \to d}(f) | L_{\tilde{q}}^{\tilde{d}} U_{\tilde{d} \to d}(g)]_d = [L_{\tilde{q}}^{\tilde{d}} U_{\tilde{d} \to d} J_{\tilde{d}}^{\tilde{q}}(F) | L_{\tilde{q}}^{\tilde{d}} U_{\tilde{d} \to d} J_{\tilde{d}}^{\tilde{q}}(G)]_d$$

if we write $f = J_{\tilde{d}}^{\tilde{q}}(F)$ and $g = J_{\tilde{d}}^{\tilde{q}}(G)$. We continue simply:

$$[L_{\tilde{q}}^{\tilde{d}} U_{\tilde{d} \to d} J_{\tilde{d}}^{\tilde{q}}(F) | L_{\tilde{q}}^{\tilde{d}} U_{\tilde{d} \to d} J_{\tilde{d}}^{\tilde{q}}(G)]_d = [U_{\tilde{q} \to d}(F) | U_{\tilde{q} \to d}(G)]_d$$

$$= [U_{\tilde{q} \to d}(F) | G]_d = [L_{\tilde{q}}^{\tilde{d}} U_{\tilde{d} \to d} J_{\tilde{d}}^{\tilde{q}}(F) | G]_d = [U_{\tilde{d} \to d} J_{\tilde{d}}^{\tilde{q}}(F) | J_{\tilde{d}}^{\tilde{q}}(G)]_d$$

$$= [U_{\tilde{d} \to d}(f) | g]_d = [U_{\tilde{d} \to d}(f) | U_{\tilde{d} \to d}(g)]_d$$

indeed proving that the restriction of $L_{\tilde{q}}^{\tilde{d}}$ to $\mathfrak{M}(\tilde{q} \to d)$ is an isometry. To show that both operators are inverses of each other, we note that

$$L_{\tilde{q}}^{\tilde{d}} \left(J_{\tilde{d}}^{\tilde{q}} L_{\tilde{q}}^{\tilde{d}} U_{\tilde{d} \to d}(f) \right) = U_{\tilde{q} \to d} U_{\tilde{q} \to d}(F) = U_{\tilde{q} \to d}(F) = L_{\tilde{q}}^{\tilde{d}} \left(U_{\tilde{d} \to d}(f) \right)$$

and since L is an injection, this indeed implies that

$$J_{\tilde{d}}^{\tilde{q}} L_{\tilde{q}}^{\tilde{d}} U_{\tilde{d} \to d}(f) = U_{\tilde{d} \to d}(f).$$

The reverse equation

$$L_{\tilde{q}}^{\tilde{d}} J_{\tilde{d}}^{\tilde{q}} U_{\tilde{q} \to d}(f') = U_{\tilde{q} \to d}(f')$$

is readily proved. $\diamond\diamond\diamond$

Thus in the relation

$$(4.11) \qquad \mathscr{F}(\mathcal{K}_q) = \overset{\perp}{\underset{r|q}{\oplus}} \mathfrak{M}(\tilde{q} \to r)$$

we may regroup $\overset{\perp}{\underset{r|d}{\oplus}} \mathfrak{M}(\tilde{q} \to r)$ for some divisor d of q and identify it with $\mathscr{F}(\mathcal{K}_d)$ via L or J, and this identification respects each summand. We may then identify $\mathscr{F}(\mathcal{K}_d)$ with the set of functions of $\mathscr{F}(\mathcal{K}_q)$ that depend only on the class of the variable modulo d, and $\mathfrak{M}(\tilde{q} \to r)$ as being the functions that depend only on the class of the variable modulo r, where r is minimal subject to this condition. Naturally, r is some kind of a *conductor*.

We may split f according to (4.11), which we term *decomposing f in Fourier components*, and this is done via

$$(4.12) \qquad f = \sum_{r|q} U_{\tilde{q} \to r}(f).$$

Note finally that

$$(4.13) \qquad \left\| U_{\tilde{q} \to r}(f) \right\|_q^2 = \frac{1}{|\mathcal{K}_q|} \sum_{n \in \mathcal{K}_q} \left| \sum_{d|q} \mu(q/d) \frac{|\mathcal{K}_d|}{|\mathcal{K}_q|} \sum_{m \equiv n[d]} f(m) \right|^2$$

where the reader will recognize the expression appearing in Theorem 2.1.

4.3. Special cases

It is customary when working with the large sieve to split sums containing $e(nb/q)$ according to the reduced form $a/q = b/d$ with $(b,d) = 1$; when such sums contain Dirichlet characters then according to the conductor of this character. We show below that such decompositions are special cases of the one exhibited in (4.11).

No restriction. When the compact set $\mathcal{K}$ is $(\mathbb{Z}/q\mathbb{Z})_q$, we have at our disposal the usual Fourier decomposition

$$(4.14) \qquad f(n) = \sum_{d|q} \sum_{a \bmod^* d} \hat{f}(q, a/d) e(na/d)$$

where

$$(4.15) \qquad \hat{f}(q, a/d) = \frac{1}{q} \sum_{n \bmod q} f(n) e(-na/d).$$

This decomposition is in fact exactly the one given by (4.12), for we readily check that

$$(4.16) \qquad U_{\tilde{q} \to d}(f)(n) = \sum_{a \bmod^* d} \hat{f}(q, a/d) e(na/d).$$

Proof. Using (4.8), we infer

$$U_{\tilde{q} \to d}(f)(n) = \sum_{r|d} \mu(d/r) \frac{r}{q} \sum_{m \equiv n[r]} f(m) = \frac{1}{q} \sum_{m \bmod q} f(m) \sum_{\substack{r|d, \\ r|m-n}} r\mu(d/r)$$

where we recognize the Ramanujan sum $c_d(n-m)$, getting

$$U_{\tilde{q} \to d}(f)(n) = \frac{1}{q} \sum_{m \bmod q} f(m) c_d(n-m)$$

$$= \frac{1}{q} \sum_{a \bmod d} \sum_{m \bmod q} f(m) e(-ma/d) e(na/d)$$

which is exactly (4.16).   ◇◇◇

Restricting to the invertible elements. When the compact set $\mathcal{K}$ is $\mathcal{U}$, we can also write

$$(4.17) \qquad f = \sum_{d|q} \sum_{\substack{\chi \bmod q \\ \chi \bmod^* d}} \hat{f}(q, \chi) \chi$$

where the expression "χ mod q and χ mod* d" represents all Dirichlet characters modulo q of conductor d and where

$$(4.18) \qquad \hat{f}(q, \chi) = \frac{1}{\phi(q)} \sum_{n \bmod^* q} f(n) \overline{\chi(n)}.$$

The reader will check that here too that

$$(4.19) \qquad U_{\tilde{q} \to d}(f) = \sum_{\substack{\chi \bmod q \\ \chi \bmod^* d}} \hat{f}(q, \chi) \chi$$

holds.

4.4. Reduction to local properties

Given a sequence $(u_n)_{n \geq 1}$ carried by $\mathcal{K}$ up to level D (see section 2.3), we consider

$$(4.20) \qquad \Delta_d(u)(n) = |\mathcal{K}_d| \sum_{m \equiv n[d]} u_m$$

which is a function of $\mathscr{F}(\mathcal{K}_d)$ provided $d \leq D$, which we assume. We have chosen this normalisation because it yields

(4.21) $$J_{\tilde{d}}^{\tilde{q}}\Delta_q = \Delta_d,$$

allowing us to use either notion. In particular, it implies

(4.22) $$\left\|U_{\tilde{q}\to d}(\Delta_q(u))\right\|_q^2 = \left\|U_{\tilde{d}\to d}(\Delta_d(u))\right\|_d^2$$

whenever $d|q$. With these notations, the L.H.S. of Theorem 2.1 reads

(4.23) $$\sum_{d \leq D} G_d(Q)\left\|U_{\tilde{d}\to d}(\Delta_d(u))\right\|_d^2$$

which was the aim of this whole chapter. We have interpreted each summand from the L.H.S. of Theorem 2.1 as a (square of a) norm of a suitable orthonormal projection of our initial function $(u_n)_{n\geq 1}$. The $\left\|U_{\tilde{d}\to d}\Delta_d\right\|$ are independent of each other and we even have a geometric interpretation for these norms. Note that our space is in fact

(4.24) $$\overset{\perp}{\underset{d \leq D}{\oplus}} \mathfrak{M}(r)$$

where one should point out a peculiarity: we do not really have a space of functions over a given set, which is why we have to go through functions in the first place. What we have is either a sequence of points in $\prod_{r \leq D} \mathfrak{M}(r)$ or, if we want to keep some geometric flavour, a sequence $(f_d)_d$ with the property that $J_{\tilde{d}}^{\tilde{q}}(f_q) = f_d$, as we checked in (4.21).

In chapter 19, we shall define a very natural adjoint for Δ_q (see (19.7)).

A different proof of Theorem 2.1. Let $U'_{\tilde{q}\to d}$ be the sequence of operators U associated with $\mathcal{K} = (\mathbb{Z}/d\mathbb{Z})_d$ and we note $\|\cdot\|'_q$ the associated norm (recall that this norm depends on the ambient compact set). We also use Δ'. We reserve $U_{\tilde{q}\to d}$ for the operators associated with $\mathcal{K}$, and exceptionally here $\|\cdot\|_{\mathcal{K}_q}$ for the relevant norm. Note that if f is in $\mathscr{F}(\mathcal{K}_d)$, then $\|f\|_q'^2 = |\mathcal{K}_q|\|f\|_{\mathcal{K}_q}^2/q$ and $\Delta'_d = d\Delta_d/|\mathcal{K}_d|$. The large sieve inequality gives a bound for

(4.25) $$\sum_{d \leq D}\left\|U'_{\tilde{d}\to d}(\Delta'_d(u))\right\|_d'^2,$$

since in this case $G_d(Q) = 1$ for every d. When we know our sequence is carried by a smaller compact set $\mathcal{K}$, we may introduce this information via the following transformation (see (4.8)):

$$\left\|U'_{\tilde{q}\to q}(\Delta'_q(u))\right\|'^2_q = \sum_{d|q}\mu(q/d)\left\|\Delta'_d(u)\right\|'^2_d = \sum_{d|q}\mu(q/d)\frac{d}{|\mathcal{K}_d|}\left\|\Delta_d(u)\right\|^2_{\mathcal{K}_d}.$$

$$= \sum_{d|q}\mu(q/d)\frac{d}{|\mathcal{K}_d|}\sum_{r|d}\left\|U_{\tilde{r}\to r}\Delta_r(u)\right\|^2_{\mathcal{K}_r}.$$

Plugging this last expression into (4.25) and rearranging some terms we reach (4.23).

5 Further arithmetical applications

5.1. On a large sieve extension of the Brun-Titchmarsh Theorem

In this section, we use the large sieve extension of the Brun-Titchmarsh inequality provided by Theorem 2.1 to detect products of two primes is arithmetic progressions. Let us consider the case of primes in $[2, N]$, of which the prime number theorem tells us there are about $N/\operatorname{Log} N$. Next select a modulus q. The Brun-Titchmarsh Theorem 2.2 implies that at least

$$(5.1) \qquad \frac{\phi(q)}{2}\left(1 - \frac{\operatorname{Log} q}{\operatorname{Log} N}\right)$$

congruence classes modulo q contains a prime $\leq N$, so roughly speaking slightly less than $\phi(q)/2$ when q is N^ε. If this cardinality is $> \phi(q)/3$, one could try to use Kneser's Theorem and derive that all invertible residue classes modulo q contain a product of three primes, but the proof gets stuck: all the primes we detect – to show the cardinality is more than $\phi(q)/3$ – could belong to a quadratic subgroup of index 2 ... However the following theorem shows that if this is indeed the case for a given modulus q then the number of classes covered modulo some q' prime to q is much larger:

Theorem 5.1. *Let $N \geq 2$. Set $\mathcal{P}$ to be the set of primes in $]\sqrt{N}, N]$, of cardinality P, and let $(q_i)_{i \in I}$ be a finite set of pairwise coprime moduli, not all more than $\sqrt{N}/\operatorname{Log} N$. Define*

$$\mathcal{A}(q_i) = \{a \in \mathbb{Z}/q_i\mathbb{Z}/ \ \exists p \in \mathcal{P}, p \equiv a[q_i]\}.$$

As N goes to infinity, we have

$$\sum_{i \in I}\left(1 - \frac{2\operatorname{Log} q_i}{\operatorname{Log} N}\right)\left(\frac{\phi(q_i)}{|\mathcal{A}(q_i)|} - 1\right) \leq 1 + o(1).$$

A similar Theorem is an essential ingredient of (Ramaré, 2007b). Let us note for historical reference that (Erdös, 1937) already showed a result of similar flavour, though weaker in several respects. See also chapter 6 for a different hindsight on the problem and Theorem 21.3 of the appendix for a similar reasoning in a more general context.

Proof. We first present a proof when q_i's are prime numbers. Applying Theorem 2.1 to the characteristic function (u_n) of the primes in $\mathcal{P}$ and the compact set $\mathcal{K} = \mathcal{U}$ and then reducing the summation to summands from (q_i), we get

(5.2)
$$G_1(Q)P^2 + \sum_{i \in I} G_{q_i}(Q)\phi(q_i) \sum_{b \bmod^* q_i} \left| \sum_{\substack{p \in \mathcal{P} \\ p \equiv b[q_i]}} 1 - P/\phi(q_i) \right|^2 \le P(N + Q^2).$$

Now applying Lemma 2.3 to get a lower bound for $G_{q_i}(Q)$ in terms of G_1, which we bound in turn by (2.13), we infer that

$$P^2 \operatorname{Log} Q + \sum_{i \in I} \operatorname{Log}(Q/q_i)\phi(q_i) \sum_{b \bmod^* q_i} \left| \sum_{\substack{p \in \mathcal{P} \\ p \equiv b[q_i]}} 1 - P/\phi(q_i) \right|^2 \le P(N + Q^2).$$

With i fixed, set

(5.3)
$$x_b = \sum_{\substack{p \in \mathcal{P} \\ p \equiv b[q_i]}} 1.$$

We know that $\sum_b x_b = P$ and that x_b is zero when b is not in $\mathcal{A}(q_i)$ and we seek the minimum of $\sum_b (x_b - P/\phi(q))^2$. This is most immediately done on setting the non-zero x_b to be all equal to $P/|\mathcal{A}(q_i)|$. A modest calculation then reveals that

(5.4)
$$\sum_{i \in I} \left(1 - \frac{\operatorname{Log} q_i}{\operatorname{Log} Q} \right) \left(\frac{\phi(q_i)}{|\mathcal{A}(q_i)|} - 1 \right) \le \frac{N + Q^2}{P \operatorname{Log} Q} - 1.$$

Setting $Q = \sqrt{N}/\operatorname{Log} N$ yields the inequality we claimed. To extend the proof to non prime moduli, we have to check that (5.2) still holds in this case, provided the q_i's are pairwise coprime. The required identity follows on combining (5.6) below together with

(5.5)
$$\sum_{d|q} |\mathcal{K}_d| \sum_{b \in \mathcal{K}_d} \left| \sum_{\ell|d} \mu(d/\ell) \frac{|\mathcal{K}_\ell|}{|\mathcal{K}_d|} \sum_{m \equiv b[\ell]} u_m \right|^2 = |\mathcal{K}_q| \sum_{b \in \mathcal{K}_q} \left| \sum_{m \equiv b[q]} u_m \right|^2$$

whenever u_m vanishes if m is not in $\mathcal{K}_q$, and where $\mathcal{K}$ is a multiplicatively split compact set verifying the Johnsen-Gallagher condition. The case $\mathcal{K} = \mathcal{U}$ would of course be enough for us here, but we can as easily get to the general case. To prove this latter identity, rewrite the L.H.S. as $\sum_{d|q} \Theta(d)$ where $\Theta(d)$ is being defined in (2.10). We showed there that

$$\Theta(d) = \sum_{r|d} \mu(d/r) |\mathcal{K}_r| \sum_{m \equiv n[r]} u_m \overline{u_n}$$

so that

$$\sum_{d|q} \Theta(d) = |\mathcal{K}_q| \sum_{m \equiv n[q]} u_m \overline{u_n}.$$

Our claim follows readily on noting that
(5.6)

$$|\mathcal{K}_q| \sum_{b \in \mathcal{K}_q} \left| \sum_{m \equiv b[q]} u_m \right|^2 - \left| \sum_m u_m \right|^2 = |\mathcal{K}_q| \sum_{b \in \mathcal{K}_q} \left| \sum_{m \equiv b[q]} u_m - \frac{\sum_m u_m}{|\mathcal{K}_q|} \right|^2.$$

An alternative proof in the case of primes can be worked out by using the expression of $\Theta(q)$ obtained in the proof of the Bombieri-Davenport Theorem, here Theorem 2.3. A third and much more conceptual proof is available to the reader who has gone through chapter 4: it proceeds by noticing that (5.5) may simply be rewritten as

$$\sum_{d|q} \left| [U_{\tilde{q} \to d}(\Delta_q(u))|\Delta_q(u)]_q \right|^2 = \|\Delta_q(u)\|_q^2$$

on joining (4.1) and (4.13) with (4.20), a relation which holds by (4.12).
◇◇◇

Corollary 5.1. *Let us consider the set of primes $\leq N$. Let q_1 and q_2 be two coprime moduli both not more than $N^{1/5}$. Then modulo q_1 or q_2, when N is large enough, all invertible residue classes contain a product of two primes.*

The limit of this corollary is $q_i \leq N^{\frac{1}{4}-\varepsilon}$. Taking three or more moduli would of course reduce this limitation.

Proof. We apply the preceding theorem to obtain that for q_1 or q_2, say for q, we have

$$2\left(1 - \frac{2\operatorname{Log} q}{\operatorname{Log} N}\right)\left(\frac{\phi(q)}{|\mathcal{A}(q)|} - 1\right) \leq 1 + o(1)$$

from which we infer that $|\mathcal{A}(q)|/\phi(q) > 1/2$. It is then classical additive number theory (applied to the multiplicative group of $\mathbb{Z}/q\mathbb{Z}$) to prove the result: for each invertible residue class b modulo q, the set $\{ba^{-1}, a \in \mathcal{A}(q)\}$ has more than $\phi(q)/2$ elements and this implies that its intersection with $\mathcal{A}(q)$ is non-empty. On considering an element in this intersection, one gets an expression of b as $a_1 a_2$ as required.   ◇◇◇

We shall use this approach in a different example in section 21.4. It may appear surprising at first sight that we should be able to find a product of two primes (exactly two primes, and not "having at most two prime factors") in an arithmetic progression to a better level than what one gets for a single prime. This is due to the additionnal structure this set has and which we put to use.

5.2. Improving on the large sieve inequality for sifted sequences

We next use Theorem 2.1 to refine the large sieve inequality.

Theorem 5.2. *Assume $\mathcal{K}$ is multiplicatively split and verifies the Johnsen-Gallagher condition (2.4). Let (u_n) be a sequence carried by $\mathcal{K}$ up to level Q. Then for $Q_0 \leq Q$, we have*

$$\sum_{q \leq Q_0} \sum_{a \bmod^* q} \left| \sum_n u_n e(na/q) \right|^2 \leq \frac{G_1(Q_0)}{G_1(Q/Q_0)} \sum_n |u_n|^2 (N + Q^2)$$

Proof. Call $\Sigma(Q_0)$ the L.H.S. of the above inequality. By Theorem 2.1 and using the notation $\Theta(q)$ that appears in its proof, we get

$$\Sigma(Q_0) = \sum_{q \leq Q_0} G_q(Q) \frac{G_q(Q_0)}{G_q(Q)} \Theta(q) \leq \max_{q \leq Q_0} \left(\frac{G_q(Q_0)}{G_q(Q)} \right) \sum_{q \leq Q_0} G_q(Q) \Theta(q)$$

$$\leq \max_{q \leq Q_0} \left(\frac{G_q(Q_0)}{G_q(Q)} \right) \sum_{q \leq Q} G_q(Q) \Theta(q) = \max_{q \leq Q_0} \left(\frac{G_q(Q_0)}{G_q(Q)} \right) \Sigma(Q)$$

from which we conclude via Lemma 2.3.   ◇◇◇

This inequality refines the large sieve inequality when Q_0 is small while Q is large (but $\leq \sqrt{N}$ in what we have in mind). Using directly the large sieve inequality for Q_0 would lose the fact that we can indeed sieve up to Q, information that we preserve in the above theorem.

Without going into any further details, let us mention that this inequality is optimal, at least in full generality and that even its specialization to the case of primes below *is* optimal, up to the constant implied in the $\ll$-symbol.

This refined inequality has been used in (Ramaré & Ruzsa, 2001).

5.3. An improved large sieve inequality for primes

We proved in section 2.4 that the G_1-function associated with $\mathcal{K} = \mathcal{U}$ verifies $G_1(Q) \geq \operatorname{Log} Q$. This is in fact the true order of magnitude, but the proof is somewhat more difficult and relies on the *convolution method*. Here is rough sketch.

Proof. We start with

$$\sum_{q\geq 1} \frac{\mu^2(q)}{\phi(q)q^s} = \zeta(s+1) \prod_{p\geq 2} \left(1 + \frac{1}{(p-1)p^{s+1}} - \frac{1}{p^{2s+1}}\right).$$

The latter series, say $H(s)$ which appears on the R.H.S. is convergent for $\Re s > -1/2$. We expand it in Dirichlet series in this half plane:

$$H(s) = \sum_{n\geq 1} \frac{h(n)}{n^s}.$$

We have

$$\frac{\mu^2(q)}{\phi(q)} = \sum_{d|q} \frac{d}{q} h(d)$$

which in turn yields

$$G_1(z) = \sum_{d\leq z} dh(d) \sum_{\substack{q\leq z \\ d|q}} \frac{1}{q} = \sum_{d\leq z} h(d) \left(\text{Log}(z/d) + \gamma + \mathcal{O}(d/z)\right).$$

We check that $|h(d)| \ll d^{-1.4}$ (this 1.4 is any number < 1.5 and the constant in the $\ll$-symbol depends on this choice), from which we infer

$$G_1(z) = \text{Log}\,z + \gamma + H'(0) + \mathcal{O}(\text{Log}\,z/z^{0.4}).$$

⬦ ⬦ ⬦

In Lemma 3.5 of (Ramaré, 1995), it is proved that

(5.7) $$G_1(z) \leq \text{Log}\,z + 1.4709 \quad (\forall z \geq 1).$$

Theorem 5.3. *If $(u_n)_{n\leq N}$ is such that u_n vanishes as soon as n has a prime factor less than $\sqrt{N}$, then*

$$\sum_{q\leq Q_0} \sum_{a\,mod^*q} \left|\sum_n u_n e(na/q)\right|^2 \leq 7\frac{N\,\text{Log}\,Q_0}{\text{Log}\,N} \sum_n |u_n|^2$$

for any $Q_0 \leq \sqrt{N}$ and provided $N \geq 100$.

Proof. Once again, we translate our hypothesis by saying that (u_n) is carried by $\mathcal{K} = \mathcal{U}$ upto level $Q = \sqrt{N}$. If $Q_0 \geq N^{3/10}$, the proof follows directly from the large sieve inequality. Else, Theorem 5.2 gives us a bound that our estimates on the G-functions translate into the required statement, the contant being majorized by

$$\frac{\alpha + 1.4709/\,\text{Log}(100)}{\frac{1}{2} - \alpha} \times 2$$

where $Q_0 = N^\alpha$ and by using (5.7) together with (2.13). A numerical application concludes.   ◇ ◇ ◇

An inequality of similar strength is stated in Lemma 6.3 of (Elliott, 1985). The reader may have difficulties in noticing the connection with our result, since, in Elliott's Lemma, the sum is restricted to prime moduli and concerns primes in progressions instead of exponential sums as here, but it is really the same mechanism that makes both proofs work.

5.4. A consequence for quadratic sequences

Here, to alleviate typographical work, we use Log_k to denote the k-th iterated logarithm.

Theorem 5.4. *For every real numbers $Q_0 \geq 10$ and N, and any sequence of complex numbers (u_n), we have*

$$\sum_{q \leq Q_0} \sum_{a \bmod^* q} \left| \sum_{n \leq N} u_n e(n^2 a/q) \right|^2 \leq c(Q_0) Q_0 \cdot (N + Q_0 g(Q_0)) \cdot \sum_{n \leq N} |u_n|^2$$

with $g(x) = \exp(20 \, \mathrm{Log}_2(3x) \, \mathrm{Log}_3(9x))$ and $c(Q_0) = 4000 \, \mathrm{Log}_2^2(3Q_0)$.

A similar result with n^2 being replaced with a fixed quadratic polynomial is easily accessible by the method given here. In between, (Gyan Prakash & Ramana, 2008) generalized greatly this result by a different method and in particular, they are able to handle the case of arbitrary polynomials (instead of only quadratic ones), provided the coefficients are integers. They are even able to handle the case of arbitrary intervals. The above is still slightly better where it applies.

Proof. We first slightly modify the proof of Theorem 5.2 : we are to majorize

$$\max_{d \leq Q_0} \{ G_d(Q_0)/G_d(Q) \}.$$

Next, we should define our set $\mathcal{K}$ and the functions G. The level of sieving is $Q = \max(N, Q_0 g(Q_0))$.

When d is squarefree we take for $\mathcal{K}_d$ the set of squares and when d is not squarefree we trivially lift $\mathcal{K}_\ell$ to $\mathbb{Z}/d\mathbb{Z}$, where ℓ is the squarefree kernel of d. This set $\mathcal{K}$ satisfies the Gallagher-Johnsen condition (2.4) and is multiplicatively split. Furthermore

$$\text{(5.8)} \quad \begin{cases} |\mathcal{K}_{p^\nu}| p^{-\nu} = \dfrac{p+1}{2p} & \text{if } p \neq 2 \ \text{and} \ \nu \geq 1, \\[2mm] |\mathcal{K}_{2^\nu}| 2^{-\nu} = 1 & \text{if } \nu \geq 1. \end{cases}$$

The associated function h vanishes on non-squarefree integers and otherwise verifies

$$h(2) = 0, \quad h(p) = \frac{p-1}{p+1} \quad \text{if } p \neq 2.$$

In this situation, we have

$$G_d(Q) = \sum_{\delta/ \, [d,\delta] \leq Q} h(\delta) = \sum_{\substack{q,r \\ (q,d)=1, r|d, \\ q \leq Q/d}} h(r) h(q)$$

i.e.

$$\text{(5.9)} \qquad G_d(Q) = \frac{d}{|\mathcal{K}_d|} \sum_{\substack{q \leq Q/d \\ (q,d)=1}} h(q)$$

(by writing $\delta = qr$) as soon as $d \leq Q$. When f is an odd integer we infer that

$$\text{(5.10)} \qquad G_{2^u f}(Q) = \prod_{p|f} \frac{2p}{p+1} \sum_{\substack{q \leq Q/(2^u f) \\ (q,2f)=1}} \mu^2(q) \prod_{p|q} \left(\frac{p-1}{p+1} \right).$$

Note that we should pay attention to the dependence in u and f while evaluating these averages. Define the multiplicative functions a and b by

$$\begin{cases} a(p) = \dfrac{p-1}{p+1} & \text{when } p \nmid 2f, \\[2mm] a(p^\nu) = 0 & \text{otherwise,} \end{cases} \qquad \begin{cases} b(p) = \dfrac{-2}{p+1} & \text{when } p \nmid 2f, \\[2mm] b(p) = -1 & \text{otherwise,} \\[2mm] b(p^2) = -\dfrac{p-1}{p+1} & \text{when } p \nmid 2f, \\[2mm] b(p^\nu) = 0 & \text{otherwise,} \end{cases}$$

so that $a = \mathbb{1} \star b$ and we have

$$G_{2^u f}(Q) = \prod_{p|f} \frac{2p}{p+1} \sum_{q \leq Q/(2^u f)} a(q).$$

We continue by appealing to Rankin's method. We use

$$\sum_{n \leq X} 1 = X + \mathcal{O}^*(X^\alpha) \qquad (\alpha > 0, X \geq 0)$$

to get

$$\sum_{q \leq D} a(q) = \sum_{d \geq 1} b(d) \left\{ \frac{D}{d} + \mathcal{O}^*((D/d)^{\alpha}) \right\}$$

$$= B(f)D + \mathcal{O}^* \left(D^{\alpha} B^* \prod_{p|2f} \left(1 + \frac{1}{p^{\alpha}} \right) \right)$$

where $B(f) = \sum_{d \geq 1} b(d)/d$ verifies

$$B(f) = \prod_{p \geq 2} \left(1 - \frac{3p-1}{p^2(p+1)} \right) \prod_{p|2f} \left(1 - \frac{1}{p} \right) \left(1 - \frac{3p-1}{p^2(p+1)} \right)^{-1}$$

$$\geq 0.35 \prod_{p|2f} \left(1 - \frac{1}{p} \right) \left(1 - \frac{3p-1}{p^2(p+1)} \right)^{-1} \left(1 + \frac{1}{p} \right) \prod_{p|2f} \left(1 + \frac{1}{p} \right)^{-1}$$

$$\geq 0.35 \prod_{p|2f} \left(1 + \frac{1}{p} \right)^{-1}.$$

We choose $\alpha = \max(3/4, 1 - 1/\operatorname{Log}_2(3f))$. Note that, for $\alpha \geq 3/4$,

$$B^* = \prod_{p \geq 3} \left(1 + \frac{2}{(p+1)p^{3/4}} + \frac{p-1}{(p+1)p^{3/2}} \right) \leq 2.3,$$

hence, by getting rid of the Euler factor at 2, we get

$$\sum_{q \leq D} a(q) = B(f)D \left(1 + \mathcal{O}^* \left(6.7 D^{\alpha-1} \prod_{p|f} \left(1 + \frac{1}{p^{\alpha}} \right)^2 \right) \right).$$

Odd integers $f \leq e^{e^4}/3$ have not more than 18 prime factors, since the product of the 18 first odd primes is greater than $e^{e^4}/3$. This implies that $\prod_{p|f}(1 + p^{-3/4})$ not more than 55 for those f's. We proceed to majorize this product when $f \geq e^{e^4}/3$. Setting $L = \operatorname{Log} 3f$, we readily check that

$$\prod_{\substack{p|f \\ p \geq L}} \left(1 + \frac{1}{p^{\alpha}} \right)^2 \leq \exp \left(\frac{2 \operatorname{Log}(f)}{L^{\alpha} \operatorname{Log} L} \right) \leq 1.2$$

since there are at most $\operatorname{Log}(f)/\operatorname{Log} L$ prime divisors of f that are $\geq L$.

On the other hand, and on using the elementary $\operatorname{Log}(1+x) - \operatorname{Log}(1+y) \leq x - y$ when $0 \leq y \leq x$, we get

$$\prod_{p \leq L} \left(1 + \frac{1}{p^{\alpha}} \right) \left(1 + \frac{1}{p} \right)^{-1} \leq \exp \sum_{p \leq L} \left(\frac{1}{p^{\alpha}} - \frac{1}{p} \right) \leq \exp \sum_{p \leq L} \frac{e^{(1-\alpha)\operatorname{Log} p} - 1}{p}.$$

We now utilize $e^x - 1 \le ex$ when $0 \le x \le 1$, getting

$$\prod_{p \le L}\left(1 + \frac{1}{p^\alpha}\right)^2\left(1 + \frac{1}{p}\right)^{-2} \le \exp\left(2.4(1-\alpha)\sum_{p \le L}\frac{\text{Log }p}{p}\right) \le e^{2.4} \le 12$$

since $\sum_{p \le L}(\text{Log }p)/p \le \text{Log }L$ by (3.24) of (Rosser & Schoenfeld, 1962). This leads to

$$\prod_{p|f,p \le L}\left(1 + \frac{1}{p^\alpha}\right)^2 \le 12\prod_{p \le L}\left(1 + \frac{1}{p}\right)^2 \le 12\exp\left(\sum_{p \le L}\frac{2}{p}\right)$$

$$\le 12\exp(2\,\text{Log Log }L + 2) \le 89\,\text{Log}_2(3f)^2$$

since $(\sum_{p \le L}1/p \le \text{Log Log }L + 1$ when $L \ge 3$ by (3.20) of (Rosser & Schoenfeld, 1962). Gathering our estimates, we infer

$$1 - 716\,\text{Log}_2(3f)^2\exp(-\text{Log }D/\text{Log}_2(3f))$$

$$\le \frac{1}{DB(f)}\sum_{q \le D}a(q) \le 716\,\text{Log}_2(3f)^2$$

when $3f \ge e^{e^4}$. The reader will easily get a better bound when f is smaller. We will use this lower estimate with $D = Q_0/(2^u f)$ and the upper one with $D = Q/(2^u f)$. Since

(5.11) $$\text{Log}(D) \ge \text{Log}(Q/Q_0) \ge 20\,\text{Log}_2(3Q_0)\,\text{Log}_3(9Q_0)$$

we get

$$1 - 716\,\text{Log}_2(3f)^2\exp(-\text{Log }D/\text{Log}_2(3f)) \ge 1 - \frac{716}{\text{Log}_2(9Q_0)^{18}} \ge 1/5.$$

Finally

$$\max_{d \le Q_0}\{G_d(Q_0)/G_d(Q)\} \le 4000(\text{Log}_2(3Q_0))^2Q_0/Q$$

ending the proof. $\diamond\diamond\diamond$

5.5. A Bombieri-Vinogradov type Theorem

We can establish Bombieri-Vinogradov type of results from Theorem 5.4 by adapting the scheme developed in (Bombieri et al., 1986). We use the notation $p \sim P$ to say $P < p \le 2P$.

Theorem 5.5. *For every $A \geq 1$, there exists B such that for all $D \leq P(\mathrm{Log}\, P)^{-B}$ and $D \leq N/g(N)$, we have*

$$\sum_{d \leq D} \max_{a \bmod^* d} \left| \sum_{\substack{n \sim N, p \sim P \\ (n^2+1)p \equiv a[d]}} \alpha(n) - \frac{\tilde{\pi}(P)}{\phi(d)} \sum_{(n^2+1,d)=1} \alpha(n) \right| \ll_A \frac{PN^{1/2}\|\alpha\|}{(\mathrm{Log}\, P)^A}$$

for any sequence of complex numbers $(\alpha(n))_n$ and where $\tilde{\pi}(P)$ is the number of primes $p \sim P$.

The function g appearing in this statement is of course the one appearing in Theorem 5.4.

For instance, this implies that the level of distribution of the sequence of $(p_1^2 + 1)p_2$ with p_1 and p_2 prime numbers such that $N \leq p_1, p_2 \leq 2N$ is larger than N. Using the general theory of the weighted sieve (see for instance (Greaves, 2001)), we infer that the sequence $1 + (p_1^2 + 1)p_2$ with $\frac{1}{2} \leq p_1/p_2 \leq 2$ contains infinitely many elements having at most four prime factors. This special result has already been proved by (Greaves, 1974); Greaves's result is more general than ours in some aspects while ours prevails in some other. For the sequence $2 + p_1^2 p_2$, it is possible to simplify the following proof by appealing to the Barban-Davenport-Halberstam Theorem.

Proof. Let us put $D = P/(\mathrm{Log}\, P)^{2A+4}$. We study

$$\Sigma = \sum_{d \leq D} \max_{a \bmod^* d} \left| \sum_{\substack{n \sim N, p \sim P \\ (n^2+1)p \equiv a[d]}} \alpha(n) - \frac{\tilde{\pi}(P)}{\phi(d)} \sum_{(n^2+1,d)=1} \alpha(n) \right|.$$

We have

$$\Sigma = \sum_{d \leq D} \max_{a \bmod^* d} \left| \frac{1}{\phi(d)} \sum_{\substack{\chi \bmod d \\ \chi \neq \chi_0}} S(\chi) T(\chi) \bar{\chi}(a) \right|$$

$$\leq \sum_{d \leq D} \frac{1}{\phi(d)} \sum_{\substack{\chi \bmod d \\ \chi \neq \chi_0}} |S(\chi)|\, |T(\chi)|$$

with

(5.12) $$S(\chi) = \sum_{n \sim N} \alpha(n)\, \chi(n^2 + 1) \quad \text{and} \quad T(\chi) = \sum_{p \sim P} \chi(p).$$

As usual, we infer that

$$\Sigma \ll \text{Log}\, D \sum_{1<q\leq D} \frac{1}{\phi(q)} \sum_{\chi \bmod {}^* q} |S(\chi)|\,|T(\chi)|$$

$$\ll (\text{Log}\, D)^2 \Big(\sum_{1<q\leq D_0} \sum_{\chi \bmod {}^* q} |S(\chi)|^2 \Big)^{1/2} \max_{\substack{\chi \bmod {}^* q \\ 1<q\leq D_0}} |T(\chi)|$$

$$+(\text{Log}\, D)\Sigma'.$$

Let us recall the classical inequality of (Gallagher, 1967)

$$(5.13) \quad \sum_{\chi \bmod {}^* q} |S(\chi)|^2 \leq \frac{\phi(q)}{q} \sum_{a \bmod {}^* q} \Big| \sum_{n \sim N} \alpha(n)\, e((n^2+1)a/q) \Big|^2.$$

Using the Siegel-Walfish Theorem (which we recall later in Lemma 10.4 in section 10.4) for $T(\chi)$ and Theorem 5.4 for $S(\chi)$ through the above inequality, we get

$$\Sigma \ll_{C_1} (\text{Log}\, P)^{-C_1} P D_0^{1/2}(\text{Log}\, D_0)(N + D_0 g(D_0))^{1/2}\|\alpha\|_2 + (\text{Log}\, D)\Sigma'$$

for $D_0 = (\text{Log}\, P)^{2A+6}$ and $C_1 = 2A + 6$. As for Σ', we split the summation over q according to the size of this parameter. We are then left with the problem of finding an upper bound for

$$\Sigma''(Q) = \frac{\text{Log}\, Q}{Q} \sum_{Q<q\leq 2Q} \sum_{\chi \bmod {}^* q} |S(\chi)|\,|T(\chi)|$$

which we treat using the Cauchy-Schwarz and the large sieve inequality :

$$\Sigma''(Q) \ll \frac{\text{Log}\, Q}{Q}(P + P^{1/2}Q) \Big(\sum_{Q<q\leq 2Q} \sum_{\chi \bmod {}^* q} |S(\chi)|^2 \Big)^{1/2}.$$

Invoking Theorem 5.4, we get for $N \geq D g(D)$

$$\Sigma''(Q) \ll \frac{\text{Log}^2 Q}{Q}(P + P^{1/2}Q)Q^{1/2}\|\alpha\|_2 N^{1/2}$$

$$\ll \|\alpha\|_2 P^{1/2} N^{1/2}\Big(\frac{P^{1/2}}{Q^{1/2}} + Q^{1/2}\Big) \text{Log}^2 Q$$

Hence

$$\Sigma \ll_A \|\alpha\|_2 (\text{Log}\, D)^3 P N^{1/2} \left((\text{Log}\, P)^{-C_1} D_0^{1/2} + \frac{1}{D_0^{1/2}} + \frac{D^{1/2}}{P^{1/2}} \right)$$

and the theorem follows readily.   ◇ ◇ ◇

6 The Siegel zero effect

When dealing with the Brun-Titchmarsh Theorem (Theorem 2.2 of this monograph), and in general, with sieve methods, the question of the connections between the parity principle, the constant 2 in this theorem and the so-called Siegel zeros cannot be avoided. (Selberg, 1949) shows that the constant $2 + o(1)$ in the Brun-Titchmarsh Theorem is optimal, *if we stick to a sieve method in a fairly general context.* He expanded this theory into what is known as the "parity principle" in (Selberg, 1972). See also (Bombieri, 1976). However, this objection is methological and belongs much more to the realm of the combinatorial sieve. In the restricted framework of the Brun-Tichmarsh Theorem, or in the even more restricted framework of this Theorem for the initial interval only, the constant 2 and "the parity principle" are indeed two different issues. This chapter is first devoted to links and parallels between Siegel zeros and the constant 2 in the aforementioned Theorem.

We complete this chapter with large sieve estimates on the number of quadratic characters χ for which the least prime p with $\chi(p) = -1$ (resp. $\chi(p) = 1$) is large.

6.1. Zeros free regions and Siegel zeros

Let us start with a Theorem initially due to de la Vallée-Poussin in 1896, which we present in the refined form given in (Kadiri, 2002):

Theorem 6.1. *The Dirichlet L-functions associated to the modulus q do not vanish in the region*

$$\Re s \geq 1 - \frac{1}{R \log(q \max(1, |\Im s|))} \quad \text{with } R = 6.4355,$$

with the exception of at most one of them. This exception correponds to a real character and has at most one real zero in the given region.

This zero is called the "Siegel zero", or sometimes the "exceptional zero". The reader should note that such a definition depends on the shape of the region, and particularly on the value of R. Some authors call "Siegel zero" a sequence of such zeros when R goes to 0. We know since Dirichlet in 1839-40 that such a zero cannot be in $s = 1$, but it can be very close to it. First there is a link between this zero and the size of $L(1, \chi)$, where χ is the associated real Dirichlet character. This link is not as tight as one could expect, but is strong enough for our purpose. The first part is a theorem due to Hecke around 1915 which can be found in (Landau, 1918). The precise form we state comes from (Pintz, 1976).

Theorem 6.2. *When an L-function belonging to the real non-principal character* χ *modulo* $q \geq 200$ *has no zero in the interval* $[1 - \alpha, 1]$, *where* $0 < \alpha < (20 \operatorname{Log} q)^{-1}$, *we have* $L(1, \chi) > 0.23\,\alpha$.

Which implies that if $L(1, \chi) = o(1/\operatorname{Log} q)$, then there is an exceptional zero. A converse statement arises from the following lemma:

Lemma 6.1. *When an L-function belonging to the real non-principal character* χ *modulo* q, *where* $q \geq 200$, *has a real zero* $\beta \geq 1 - (\operatorname{Log} q)^{-1}$, *then* $L(1, \chi) \leq 2(1 - \beta) \operatorname{Log}^2 q$.

Proof. The mean value Theorem tells us that there exists a u in $[\beta, 1]$ such that $L(1, \chi) - 0 = (1 - \beta)L'(u, \chi)$. We bound the latter quantity trivially:

$$L'(u, \chi) = \sum_{n \leq q} \frac{\chi(n) \operatorname{Log} n}{n^u} + \int_q^\infty \sum_{q < n \leq t} \chi(n) \frac{u \operatorname{Log} t - 1}{t^{u+1}} dt$$

and hence

$$|L'(u, \chi)| \leq e \left(\frac{\operatorname{Log}^2 q}{2} - \frac{\operatorname{Log}^2 3}{2} + \frac{\operatorname{Log} 3}{3} + \frac{\operatorname{Log} 2}{2} \right) + q \frac{\operatorname{Log} q}{q^u}$$

$$\leq e \left(\frac{\operatorname{Log}^2 q}{2} + \operatorname{Log} q + 0.11 \right) \leq 2 \operatorname{Log}^2 q.$$

◇ ◇ ◇

See also (Goldfeld & Schinzel, 1975), as well as the mentioned paper of Pintz for more precise links between $L(1, \chi)$ and $1 - \beta$.

(Landau, 1918) proved that the modulus associated to any two such zeros cannot be close one to the other. Here is the latest result due to (Kadiri, 2007) in this direction:

Theorem 6.3. *Let* χ_1 *modulo* q_1 *and* χ_2 *modulo* q_2 *be two real primitive characters, and let* $\beta_1 > 0$ *(resp.* $\beta_2 > 0$) *be a real zero of* $L(s, \chi_1)$ *(resp.* $L(s, \chi_2)$). *Assume that* q_1 *and* q_2 *are coprime. Then*

$$\min(\beta_1, \beta_2) \leq 1 - \frac{1}{2.31 \operatorname{Log}(q_1 q_2 / 47)}.$$

The reader may wonder why such zeros are called *Siegel* zeros, and indeed the name *Landau-Siegel* zeros may well be better suited, since

(Landau, 1935) is the very first success at proving a result like Theorem 6.4 below with an $\varepsilon < 1/2$ (Landau still required $\varepsilon > 3/8$). Here is a version from (Tatuzawa, 1951) of the Theorem of (Siegel, 1935) that warranted this nomenclature.

Theorem 6.4. *For any $\varepsilon > 0$, and any primitive real character χ modulo q, we have $L(1, \chi) \geq \varepsilon/(10q^\varepsilon)$ with the exception of at most one value of q.*

The reader may consult (Hoffstein, 1980) as well as (Ji & Lu, 2004). for improved versions of this result. Theorem 6.4 has the following consequence: for any $\varepsilon > 0$, there exists a constant $c(\varepsilon) > 0$ such that $L(1, \chi) > c(\varepsilon)q^{-\varepsilon}$. However, this proof does not allow us to effectively compute the constant $c(\varepsilon)$, even if we take $\varepsilon = 1/3$ for instance. As a matter of fact, we know an effective solution only in the case $\varepsilon = 1/2$.

On this subject, the reader may read the groundbreaking paper of (Goldfeld, 1985) as well as (Gross & Zagier, 1983) and (Oesterlé, 1985).

6.2. Gallagher's prime number Theorem

The existence of a possible exceptional zero has a deep impact on the distribution of primes in arithmetic progressions. The theorem we present here is one of the finest achievements in this direction and clarifies greatly the situation. Some of the results we seek can be shown without having to appeal to such a heavyweight, but using it is enlightening.

The prime number theorem of (Gallagher, 1970) has a long ancestry, steming originally from (Linnik, 1944a) and (Linnik, 1944b). Another modern form of these celebrated papers can be found in (Bombieri, 1987), Theorem 16. See also (Motohashi, 1978).

One of the key to such results is the Deuring-Heilbronn phenomenon: when there is an exceptional zero, all other L-functions have no zero in a region wider than usual, and this region becomes wider as this exceptional zero closes to 1.

Let us now state Gallagher's Theorem. Assume $L(\beta, \chi) = 0$ for a β such that $1 - \beta = o(1/\operatorname{Log} q)$. We set $\delta = 1 - \beta$. In this case

$$(6.1) \quad \psi(X; q, \ell) = \frac{X}{\phi(q)} \left(1 - \chi(\ell) \frac{X^{-\delta}}{\beta} \right)$$

$$+ \mathcal{O}\left(\frac{X\delta \operatorname{Log} T}{\phi(q)} \left(\frac{e^{-c_1 \frac{\operatorname{Log} X}{\operatorname{Log} T}}}{\operatorname{Log} X} + \frac{q \operatorname{Log} X}{\sqrt{T}} + \frac{T^{5.5}}{\sqrt{x}} \right) \right)$$

if $X \geq T^{c_2} \geq T \geq q$ where $c_1, c_2 > 0$ are two effective constants. The constant implied in the $\mathcal{O}$-symbol is equally effective. If no exceptional zero exists modulo q (that is, also for no divisor of q), the preceding formula holds with minor modifications: we use $\beta = \frac{1}{2}$ in the main term and $\delta \operatorname{Log} T = 1$ in the remainder term.

6.3. Siegel zero and Brun-Titchmarsh Theorem

We prove here the following Theorem whose idea comes from (Motohashi, 1979), where a similar result is proved by a very different method.

Theorem 6.5. *There exist two effective constants c_3 and c_4, such that for $q \geq c_4$, the following two conditions are equivalent.*

  (1) For any real character modulo q, we have $L(1, \chi) \gg 1/\operatorname{Log} q$.
  (2) There exist a constant $\xi > 0$ such that for any ℓ prime to q, we have, with $X = q^{c_3}$:

$$(6.2) \qquad \sum_{\substack{X < p \leq 2X, \\ p \equiv \ell[q]}} 1 \leq \frac{2 - \xi}{\phi(q)} \sum_{X < p \leq 2X} 1.$$

Such a statement is always somewhat tricky. For instance, we indeed use characters and not only primitive characters. We can take $c_3 = \max(36, 3c_2)$, where c_2 appears in (6.1).

Proof. We shall use Gallagher's Theorem (6.1) with $T = q^3$ and $X \geq q^{\max(36, 3c_2)}$, so that the error term there is $\mathcal{O}$ of $X/(\phi(q) \operatorname{Log} q)$. First, assume $L(1, \chi) = o(1/\operatorname{Log} q)$ for one character χ. Then, by Hecke's theorem, there is indeed an exceptional zero, say β, associated to a character χ. We have $X^{-\delta} = 1 + o(1)$. In particular, if we take an invertible residue class ℓ such that $\chi(\ell) = -1$, we have $\psi(X; q, \ell) \sim 2X/\phi(q)$, and this readily implies that ξ cannot exist.

For the reverse implication, we follow (Ramachandra *et al.*, 1996). By summing our upper bound over all ℓ such that $\chi(\ell) = -1$, we discover that the number of primes in $]X, 2X]$ with $\chi(p) = 1$ is at least

$$\xi \sum_{X < p \leq 2X} 1/2.$$

Consider next $G(s) = \zeta(s)L(s, \chi) = \sum_{n \geq 1} g(n)n^{-s}$ where $g(n) = \mathbb{1} \star \chi(n)$. Note that $g(n)$ is non-negative. Note, furthermore, that $g(p) = 2$ when $\chi(p) = 1$, from which we infer

$$\sum_{X < n \leq 2X} g(n) \geq \sum_{X < p \leq 2X} g(p) \gg \xi X/\operatorname{Log} X.$$

This readily yields

$$\frac{1}{2i\pi} \int_{c-i\infty}^{c+i\infty} G(s+1)\Gamma(s)\big((2X)^s - X^s\big)ds$$

$$= \sum_{n\geq 1} \frac{g(n)}{n} \big(e^{-n/(2X)} - e^{-n/X}\big) \gg \sum_{X < n \leq 2X} \frac{g(n)}{X} \gg \xi/\operatorname{Log} X.$$

Next, shifting the path of integration in the above integral to $\Re s = -1/4$, we see that it is

$$L(1,\chi)\operatorname{Log} 2 + \mathcal{O}\left(X^{-1/4} \int_{c-i\infty}^{c+i\infty} |G(s+1)\Gamma(s)|ds\right).$$

The exponential decay of $\Gamma(s)$ in vertical strips (a consequence of the Stirling formula) as well as the bound $|G(3/4+it)| \ll q^{1/4}(1+|t|)$ ensures us that this last error term is at most $\mathcal{O}((q/X)^{1/4})$, which in turn is $\mathcal{O}(q^{-1})$ since $c_3 \geq 5$. So that we find that $L(1,\chi) \gg 1/\operatorname{Log} X \gg 1/\operatorname{Log} q$ as required. ◇◇◇

Thus, improving on the constant 2 in the Brun-Titchmarsh Theorem when X is a power of q would remove any Siegel zero. Note that we use only the Brun-Titchmarsh Theorem for the initial range. Drawing on similar ideas, (Basquin, 2006) established a theorem linking an effective lower bound for $L(1,\chi)$ of the shape $1/q^c$ for some $c \in]0, 1/2]$ with the improvement on the constant 2 in the Brun-Titchmarsh Theorem, but in a different range for X:

Theorem 6.6. *Let $c > 0$ be a parameter. The following three problems are equivalent:*

(1) *For every $\varepsilon > 0$, and every real character χ, prove in an effective way that $L(1,\chi) \gg q^{-c-\varepsilon}$.*

(2) *For every $\varepsilon > 0$, prove (6.2) for every $q \leq (\operatorname{Log} X)^{(1/c)-\varepsilon}$.*

(3) *For every $\varepsilon > 0$, prove in an effective way that $\psi(X; q, \ell) \sim X/\phi(q)$ for every $q \leq (\operatorname{Log} X)^{(1/c)-\varepsilon}$.*

This statement also tells us that, if we are able to beat the factor 2 in the upper bound, then a much stronger conclusion follows, namely an equivalent for $\psi(X; q, \ell)$. This situation is similar to what happens with the elementary proof of the prime number theorem, a proof this time heavily linked to the *parity principle*. See (Selberg, 1949b), (Selberg, 1949a) and (Erdös, 1949).

6.4. The Siegel zero effect

We have seen that the distribution of primes in arithmetic progressions modulo q stumbles on the possible existence of the so called Siegel zero. The existence of such a zero would have the effect that only about half the residue classes would contain primes. However, the reader should notice that this philosophical statement is sustained by theorems only when q is a small power of X.

When approaching the problem of this distribution through zeros of L-functions, this effect is well controlled and is avoided by a simple fact: two moduli q_1 and q_2 coprime and not too far apart in size cannot simultaneously have a Siegel zero by Theorem 6.3. For instance, this remedy is used in (McCurley, 1984) and (Cook, 1984). The condition of coprimality is not minor in any sense: if q has a Siegel zero, then the distribution of primes modulo $3q$ for instance is still going to be unbalanced.

From the sieve point of view, zeros do not appear as such, but a similar role is played by the fact that we can only prove that the number of primes in a given arithmetic progression is about twice what it should be. Indeed, this implies that, then, primes cannot accumulate on a subset of $(\mathbb{Z}/q\mathbb{Z})^*$ that contains less than $(1-\varepsilon)\phi(q)/2$ elements. Again, this is true only when q is small when compared to X, but, when q is larger, we can still prove that a subset of positive density (with respect to $(\mathbb{Z}/q\mathbb{Z})^*$) is attained.

We also have a similar effect to Landau's, even if we are not actually able to produce a corresponding zero. And, indeed, by using a large sieve extension of the Brun-Titchmarsh inequality, we saw in Theorem 5.1 that primes cannot accumulate in some small sets modulo two coprime moduli of similar size. Further the density of the set attained can even be shown to be rather close to 1 if we are ready to chose one modulus among say T candidates. Exactly how large depends on the size of the modulus, say q and of T, but we can roughly show that more than $\left(1 + 2\operatorname{Log} X/(T\operatorname{Log}(X/q^2))\right)^{-1}\phi(q)$ classes are reached and this indeed will be larger than a half provided T is large enough depending on $\operatorname{Log}(X/q^2)$.

This is what we loosely call the *Siegel zero effect*, though no zeros are involved. And since it finds its justification in sieves, it can be used on other sequences as well; we provide such an example in Theorem 21.3 of chapter 21.

6.5. A detour: the precursory theorem of Linnik

Proving that $L(s, \chi)$ has no zero close to 1 has to do with proving that $L(1, \chi)$ cannot be small, which means, when χ is quadratic, proving that χ often takes the value 1. Curiously enough, we do not know how to prove either that $\chi(p)$ often takes the value -1 [1], or that it takes often the value 1 [2], where here it is necessary to specify that we seek the value at prime argument for the problem to be non trivial. One of the first arithmetical use of the large sieve technique occurred in (Linnik, 1942), where the author proves

Theorem 6.7. *For every $\varepsilon > 0$, there exists $c(\varepsilon)$ such that, for every N, the number of prime numbers $\leq N$ that have no non-quadratic residue $\leq N^\varepsilon$ is at most $c(\varepsilon)$.*

We refer the reader to (Montgomery, 1971) for a more thorough treatment of the history of the subject. We now present a proof of this result. As usual we shall have to compute a density, for which we rely on the following lemma.

Lemma 6.2.
   The number of integers $\leq N$ whose prime factors are all $\leq N^\varepsilon$ is $\gg_\varepsilon N$.

There exist better proofs than the one we give now, and it is known in particular that this set has a cardinality equivalent to a constant (depending on ε) times N. However, the one we present relies once more on the idea of (Levin & Fainleib, 1967). Moreover, it appears to be novel.

Proof. Set $\epsilon = 1/k$, where $k \geq 1$ is an integer. Let $\mathcal{S}$ be the set of integers that have no prime factors $\leq N^\epsilon$ and let Z be the number of them that are $\leq N$. Let us first write

$$Z \operatorname{Log} N = \sum_{\substack{n \in \mathcal{S}, \\ n \leq N}} \operatorname{Log} N \geq \sum_{\substack{n \in \mathcal{S}, \ p \mid n \\ n \leq N}} \sum \operatorname{Log} p \geq \sum_{p \leq N^\epsilon} \operatorname{Log} p \sum_{\substack{n \in \mathcal{S}, \\ np \leq N}} 1$$

$$\geq \sum_{\substack{n \in \mathcal{S}, \\ N^{1-\epsilon} < n \leq N}} \sum_{p \leq N/n} \operatorname{Log} p \geq C_6 N \sum_{\substack{n \in \mathcal{S}, \\ N^{1-\epsilon} < n \leq N}} 1/n - C_7 Z$$

---

[1]If the conductor, say $\mathfrak{f}$, of χ is prime, then p is a non quadratic residue, i.e. is not a square in $\mathbb{Z}/\mathfrak{f}\mathbb{Z}$.

[2]This time, when the conductor of χ is prime, p would be a quadratic residue.

for some constants $C_6, C_7 > 0$. We shall get a lower bound for the sum of $1/n$ when n ranges $\mathcal{S}$ and in above interval by following a similar path. We will achieve this by a recursion whose main ingredient is the following fact: There exist two constants $c_1 = c_1(\epsilon)$ et $N_0 = N_0(\epsilon)$ such that for every $\ell \in \{0, \ldots, k-1\}$ and $N \geq N_0$, we have

$$(6.3) \qquad \sum_{\substack{n \in \mathcal{S}, \\ 2^\ell N^{1-(\ell+1)/k} < n \leq N^{1-\ell/k}}} 1/n \geq c_1 \sum_{\substack{n \in \mathcal{S}, \\ 2^{\ell+1} N^{1-(\ell+2)/k} < n \leq N^{1-(\ell+1)/k}}} 1/n.$$

Let us first establish this inequality. We write

$$\text{Log } N \sum_{\substack{n \in \mathcal{S}, \\ 2^\ell N^{1-(\ell+1)/k} < n \leq N^{1-\ell/k}}} 1/n \geq \sum_{\substack{n \in \mathcal{S}, \\ 2^\ell N^{1-(\ell+1)/k} < n \leq N^{1-\ell/k}}} (\text{Log } n)/n$$

$$\geq \sum_{\substack{n \in \mathcal{S}, \\ 2^\ell N^{1-(\ell+1)/k} < n \leq N^{1-\ell/k}}} \left(\sum_{p|n} \text{Log } p \right)/n$$

$$\geq \sum_{p \leq N^\epsilon} \frac{\text{Log } p}{p} \sum_{\substack{m \in \mathcal{S}, \\ 2^\ell N^{1-(\ell+1)/k} < mp \leq N^{1-\ell/k}}} 1/m$$

and an interchange of summations yields the lower bound

$$\sum_{\substack{m \in \mathcal{S}, \\ 2^{\ell+1} N^{1-(\ell+2)/k} < m \leq N^{1-\ell/k}}} (1/m) \sum_{\substack{p \leq N^\epsilon, \\ N^{1-(\ell+1)/k} < mp \leq N^{1-\ell/k}}} \frac{\text{Log } p}{p}.$$

If $m \leq N^{1-(\ell+1)/k}$, then the only upper bound for p is $p \leq N^\epsilon$. The lower bound reads $N^{1-(\ell+1)/k} < mp$, and

$$\sum_{\substack{p \leq N^\epsilon, \\ N^{1-(\ell+1)/k}/m < p}} \frac{\text{Log } p}{p} \geq C_1 \text{Log}(mN^{-1+(\ell+2)/k}) - C_2$$

for some constants $C_1, C_2 > 0$. When $m > N^{1-(\ell+1)/k}$, the only lower bound for p is 2, but its upper bound this time depends on m. We get

$$\sum_{p \leq N^{1-\ell/k}/m} \frac{\text{Log } p}{p} \geq C_3 \text{Log}(N^{1-\ell/k}/m) - C_4$$

for some constants $C_3, C_4 > 0$. On the other hand, when m verifies $2^{\ell+1} N^{1-(\ell+2)/k} < m \leq N^{1-(\ell+1)/k}$, then for all $p \in [\frac{1}{2}N^\epsilon, N^\epsilon]$, we have

$N^{1-(\ell+1)/k} < m' = mp \le N^{1-\ell/k}$. Since there exists $C_5 > 0$ (independent of $x \ge 2$!) such that $\sum_{x/2 \le p \le x} 1/p \ge C_5$, we reach

$$C_1 \operatorname{Log}(mN^{-1+(\ell+2)/k}) - C_2 + \sum_{\frac{1}{2}N^\epsilon \le p \le N^\epsilon} \frac{1}{p}\left(C_3 \operatorname{Log}(N^{1-\ell/k}/m') - C_4\right)$$

$$\ge C_1 \operatorname{Log}(mN^{-1+(\ell+2)/k}) - C_2 + C_5\left(C_3 \operatorname{Log}(N^{1-(\ell+1)/k}/m) - C_4\right)$$

$$\ge \min(C_1, C_5 C_3) \operatorname{Log}(N^{1/k}) - C_2 - C_5 C_4 \gg \operatorname{Log} N$$

if $N \ge N_0(\epsilon)$. We simply collect our estimates together to establish (6.3). A repeated use of it yields

$$\sum_{\substack{n \in \mathcal{S}, \\ N^{1-1/k} < n \le N}} 1/n \ge c_1^{k-1} \sum_{\substack{n \in \mathcal{S}, \\ 2^{k-1} < n \le N^{1/k}}} 1/n.$$

which is $\gg \operatorname{Log} N^\epsilon$ since the condition $n \in \mathcal{S}$ is there superfluous. Hence

$$(C_6 \operatorname{Log} N + C_7)Z \gg_\epsilon N \operatorname{Log} N$$

which is what we wished to prove. $\qquad \diamond\diamond\diamond$

Proof of Theorem 6.7. Let $\mathcal{P}$ be the set of prime numbers $\le Q = N^{1/4}$ that have no quadratic non-residue $\le N^\varepsilon$, and let $\mathcal{S}$ be the set of integers whose prime factors are $\le N^\varepsilon$. The compact set we use is defined in the following way: $\mathcal{K}_p$ is the set of quadratic residues if $p \in \mathcal{P}$, of cardinality $(p+1)/2$. If $p \notin \mathcal{P}$, we take simply $\mathcal{K}_p = \mathbb{Z}/p\mathbb{Z}$. We extend this definition to $\mathcal{K}_{p^\nu}$ by taking the inverse image of $\mathcal{K}_p$ under the canonical surjection when $\nu \ge 2$. We get a squarefree compact set. In order to apply Gallagher's theorem we first check that

$$G(Q) \ge \sum_{\substack{p \in \mathcal{P}, \\ p \le Q}} \frac{p-1}{p+1} \ge \#\{p \le Q, p \in \mathcal{P}\}/3$$

and thus

$$Z \le (N + Q^2)/G(Q)$$

which yields $G(Q) \le (N + Q^2)/Z$. Next, we notice that in the points counted in Z, we find all the integers n whose prime factors are less than N^ε: indeed each of its prime factor belongs to $\mathcal{K}$, which implies that n belongs to it also. Here we use the fact that we are looking for a quadratic residue; a similar proof would not work in the case of quadratic non-residues. As a consequence, $Z \ge c(\varepsilon)N$ for a constant $c(\varepsilon) > 0$, which in turns implies that the number of elements of $\mathcal{P}$ that are not more than $Q = N^{1/4}$ is finite. But what if we go upto N ? We simply use this result with N^4 instead of N and $\varepsilon/4$ instead of ε. $\diamond\diamond\diamond$

And what if we were to consider non-quadratic residues modulo non prime q ? If $q = q_1 q_2$ where q_1 and q_2 are coprime, and every integer $\leq N^\varepsilon$ is a square modulo q, then the same property holds also for q_1 and q_2. Let us restrict the problem to squarefree moduli q. Start from a set $\mathcal{S}$ of S moduli q such that every integer $\leq N^\varepsilon$ is a quadratic residue. The set $\mathcal{P}$ of prime divisors of every elements of $\mathcal{S}$ contains at least $(\mathrm{Log}\, S)/\mathrm{Log}\, 2$ elements and is bounded by the theorem above. This implies that S is also bounded.

6.6. And what about quadratic residues ?

The situation concerning prime quadratic residues is much less satisfactory and we are not able to prove that there exist such a prime less than the conductor, even if we are to admit a finite number of exceptions! (Elliott, 1983) and (Puchta, 2003) prove results in this direction. In this precise case, they are all a consequence of the Bombieri-Davenport Theorem 2.3. Let $\varepsilon > 0$ be given. Consider the set $\mathcal{Q}$ of moduli $q \leq Q = N^{(1/2)-\varepsilon}$ and such that there exists a primitive real character χ_q satisfying

$$\forall p \leq N, \quad \chi_q(p) = -1.$$

We take for (u_n) the characteristic function of those primes in $[\sqrt{N}, N]$ and use Theorem 2.3. We get

$$\mathrm{Log}\, \sqrt{N} \Big| \sum_{\sqrt{N} \leq p \leq N} 1 \Big|^2 + \sum_{q \in \mathcal{Q}} \mathrm{Log}(\sqrt{N}/q) \Big| \sum_{\sqrt{N} \leq p \leq N} \chi_q(p) \Big|^2 \leq 2N \sum_{\sqrt{N} \leq p \leq N} 1.$$

After some shuffling, we conclude that $|\mathcal{Q}| \ll 1/\varepsilon$. Hence, apart from a finite number of exceptions, for every primitive real character modulo $q \leq Q$, there is a prime $p \leq Q^{2+\varepsilon}$ such that $\chi_q(p) = 1$.

Note here that a smaller bound (namely $Q^{1+\varepsilon}$ instead of $Q^{2+\varepsilon}$) follows from the beautiful result of (Heath-Brown, 1995), though with a larger set of exceptions. We state this result for completeness.

Theorem 6.8. *Let $\mathcal{X}(Q)$ be the set of primitive quadratic characters of conductor $\leq Q$. Then for every $\varepsilon > 0$, we have*

$$\sum_{\chi \in \mathcal{X}(Q)} \Big| \sum_n{}^{\flat} u_n \chi(n) \Big|^2 \ll_\varepsilon (NQ)^\varepsilon (N+Q) \sum_n |u_n|^2$$

where $\sum^{\flat}$ denotes a summation restricted to squarefee integers.

7 A weighted hermitian inequality

We continue to develop the theory in the general context of chapter 1 with a view to an application in the chapter that follows.

Sometimes, a partial treatment of the bilinear form is readily available in the form of

$$(7.1) \qquad \forall(\xi_i)_i \in \mathbb{C}^I, \quad \left\| \sum_i \xi_i \varphi_i^* \right\|^2 \leq \sum_i M_i |\xi_i|^2 + \left(\sum_i |\xi_i| n_i \right)^2$$

for some positive M_i, and n_i (here again, M_i is generally an approximation to $\|\varphi_i^*\|^2$). This leads, naturally, to the definition of *a mixed almost orthogonal system*. With such an inequality at hand, the proof of Lemma 1.2 leads to the inequality

$$(7.2) \qquad \|f\|^2 - 2\Re \sum_i \overline{\xi_i}[f|\varphi_i^*] + \sum_i M_i |\xi_i|^2 + \left(\sum_i |\xi_i| n_i \right)^2 \geq 0.$$

When using it, we shall take for φ_i^* a "local approximation" to f in a sense to be made precise later on, but it already implies we can assume $[f|\varphi_i^*]$ to be a non-negative real number. It is also readily seen that the ξ_i's minimizing the R.H.S. are non-negative. Finally, we are led to choosing these ξ_i's so as to minimize

$$\|f\|^2 - 2\sum_i \xi_i[f|\varphi_i^*] + \sum_i M_i \xi_i^2 + \left(\sum_i \xi_i n_i \right)^2.$$

We handle this optimization using calculus by setting $\xi_i = \zeta_i^2$. Easy manipulations then allow us to conclude that there exists a subset I' of I such that $\xi_i = 0$ if $i \in I \setminus I'$ and

$$(7.3) \qquad \forall i \in I', \quad \xi_i = \frac{[f|\varphi_i^*] - X n_i}{M_i}, \quad X = \frac{\sum_{j \in I'} n_j [f|\varphi_j^*]/m_j}{1 + \sum_{j \in I'} n_j^2/m_j}$$

provided that

$$(7.4) \qquad \forall i \in I', \quad [f|\varphi_i^*]/n_i \geq X.$$

With these choices and hypotheses, we infer the bound

$$(7.5) \qquad \|f\|^2 + X^2 \left(1 + \sum_{j \in I'} n_j^2/m_j \right) \geq \sum_{i \in I'} M_i^{-1} |[f|\varphi_i^*]|^2.$$

However, determining optimal I' is difficult: the condition (7.4) is complicated by the appearance of the contribution from the index i on both sides. It is easier to set

$$(7.6) \qquad \xi_i = \frac{[f|\varphi_i^*] - Yn_i}{M_i},$$

for a Y to be chosen but which guarantees $\xi_i \geq 0$. The optimal Y is of course $Y = X$. Next we note that we could add a general innocuous term $\sum_{i,j} \xi_i \overline{\xi_j} \omega_{i,j}$ to (7.1) and still follow the above reasoning. Continuing in this direction, we see that it is enough to start from (1.1), but to choose the weight ξ_i given by (7.6), where this time the n_i's are to be chosen! Of course the above discussion tells the user when to use such weights, how to choose the n_i's as well as which set of moduli to select (namely take the indices i such that $\xi_i \geq 0$).

Here is the theorem we have reached:

Theorem 7.1. *Suppose that we are given an almost orthogonal system in the notations of definition 1.1. Let f be an element of $\mathcal{H}$ and Y be a real number ≥ 0. Let $(n_i)_i$ be a collection of complex numbers. Set $\xi_i = ([f|\varphi_i^*] - Yn_i)/M_i$ for each i. Then we have that*

$$\sum_i M_i |\xi_i|^2 + 2Y\Re \sum_i n_i \overline{\xi_i} - \sum_{i,j} \xi_i \overline{\xi_j} \omega_{i,j} \leq \|f\|^2$$

With $n_i = 0$, this is lemma 1.2.

8 A first use of local models

We now turn towards another way of using the large sieve inequality in an arithmetical way, here on prime numbers. This application comes from (Ramaré & Schlage-Puchta, 2008). A exposition in the French addressing a large audience can be found in (Ramaré, 2005).

8.1. Improving on the Brun-Titchmarsh Theorem

We prove the following result:

Theorem 8.1. *There exists an N_0 such that for all $N \geq N_0$ and all $M \geq 1$ we have*

$$\pi(M + N) - \pi(M) \leq \frac{2N}{\text{Log } N + 3}.$$

As we remarked earlier, (Selberg, 1949) shows that the constant $2 + o(1)$ in the above numerator is optimal, if we are to stick to a sieve method in a fairly general context.

It is thus of interest to try to quantify the $o(1)$ in $2 + o(1)$. The first upper bound of the shape $2N/(\text{Log } N + c)$ with an unspecified but very negative c is due to (van Lint & Richert, 1965) though (Selberg, 1949) mentions such a result around equation (6) of this paper, albeit without giving a proof. (Bombieri, 1971) gave the value $c = -3$ and (Montgomery & Vaughan, 1973) the value $c = 5/6$. In section 22 of "lectures on sieves", (Selberg, 1991) gives a proof for $c = 2.81$, a proof from which we have taken several elements. The treatment we present here leads to a value of c that is slightly larger than 3; it is further developed in (Ramaré & Schlage-Puchta, 2008) where the value $c = 3.53$ is obtained.

In our problem, we select an integer $\mathfrak{f}$ that will be taken to be 210 at the end of the proof and consider the characteristic function w of the points in $[M + 1, M + N]$ that are coprime with $\mathfrak{f}$. This being chosen, our scalar product on sequences over $[M + 1, M + N]$ is defined by

$$(8.1) \qquad [g|h] = \sum_{M+1 \leq n \leq M+N} w(n)g(n)\overline{h(n)}.$$

We need very refined estimates concerning this scalar product, and this is the subject of next section.

We write $\rho = \phi(\mathfrak{f})/\mathfrak{f}$ to simplify the typography.

8.2. Integers coprime to a fixed modulus in an interval

We study here the quantities

$$\begin{cases} \theta_{\mathfrak{f}}^{-}(u) = \min_{\substack{y\in\mathbb{R} \\ x\in\mathbb{R}}} \min_{0\le x\le u} \left(\sum_{\substack{y<n\le y+x, \\ (n,\mathfrak{f})=1}} 1-\rho x \right), \\[2em] \theta_{\mathfrak{f}}^{+}(u) = \max_{\substack{y\in\mathbb{R} \\ x\in\mathbb{R}}} \max_{0\le x\le u} \left(\sum_{\substack{y<n\le y+x, \\ (n,\mathfrak{f})=1}} 1-\rho x \right). \end{cases}$$

In order to compute these functions, we need to restrict both x and y to integer values. This is the role of next lemma.

Lemma 8.1. *We have*

$$\begin{cases} \theta_{\mathfrak{f}}^{-}(u) = \min_{\ell\in\mathbb{N}} \left(\min_{\substack{k\in\mathbb{N}, \\ 0\le k\le u}} \left(\sum_{\substack{\ell+1\le n\le \ell+k-1, \\ (n,\mathfrak{f})=1}} 1-\rho k \right), \sum_{\substack{\ell+1\le n\le \ell+[u], \\ (n,\mathfrak{f})=1}} 1-\rho u \right), \\[2em] \theta_{\mathfrak{f}}^{+}(u) = \max_{\substack{k,\ell\in\mathbb{N}, \\ k<u+1}} \left(\sum_{\substack{\ell\le n\le \ell+k-1, \\ (n,\mathfrak{f})=1}} 1-\rho(k-1) \right) \end{cases}$$

The function $\theta_{\mathfrak{f}}^{+}$ is a non decreasing step function which is left continuous with jumps at integer points. The function $\theta_{\mathfrak{f}}^{-}$ is non-increasing continuous : it alternates from linear pieces with slope $-\rho$ to constant pieces. The changes occur at integer points. Both are constant if $u\ge\mathfrak{f}$.

Proof. We start with $\theta_{\mathfrak{f}}^{+}$. First fix y. The function $\sum_{y<n\le y+x} w(n)-\rho x$ is linear non-increasing in x from 0 to $1-\{y\}$, then from $1-\{y\}$ to $2-\{y\}$ and so on. Its maximum value is reached at $x=0$ or $x=k-\{y\}$ for some integer k, thus

$$\theta_{\mathfrak{f}}^{+}(u) = \max_{y\in\mathbb{R}} \max_{\substack{k\in\mathbb{N}, \\ k\le u+\{y\}}} \left(\sum_{y<n\le [y]+k} w(n) + \rho(-k+\{y\}) \right).$$

The condition is increasing in $\{y\}$ and so is the term that is to be maximized. We may take y to be just below an integer ℓ, reaching the expression we announced.

Let us now consider $\theta_{\mathfrak{f}}^{-}$. We start similarly by fixing y. The minimum is reached at $x=k-\{y\}-0$ or at $x=u$, where k is an integer and the -0 means we are to take x just below this value. We get $\theta_{\mathfrak{f}}^{-}(u)$ equals to

$$\min_{y\in\mathbb{R}}\left(\min_{\substack{k\in\mathbb{N},\\ k\leq u+\{u\}}}\left(\sum_{y<n\leq[y]+k-1}w(n)+\rho(\{y\}-k)\right),\ \sum_{y<n\leq[y]+u}w(n)-\rho u\right).$$

As for the last sum, the worst case is when y is an integer $\ell\geq 0$, getting

$$(8.2)\qquad\qquad\min_{\ell\in\mathbb{N}}\left(\sum_{\ell+1\leq n\leq\ell+u}w(n)-\rho u\right).$$

For the first minimum, we distinguish between $k\leq[u]$ and $k=[u]+1$ (which can happen only if u is *not* an integer). If $k\leq[u]$, we may take y to be integral. If $k=[u]+1$, then $\{y\}\geq 1-\{u\}$ which is indeed the worst case: we take $y=\ell+1-\{u\}$. This last contribution turns out to be exactly the same as the one in (8.2).   ◇◇◇

Next we consider the function

$$(8.3)\qquad\qquad\theta_{\mathfrak{f}}^{*}(v)=\max(\theta_{\mathfrak{f}}^{+}(1/v),-\theta_{\mathfrak{f}}^{-}(1/v))$$

which this time is right continuous with jump points at $1/m$, where m ranges integers from 1 to $\mathfrak{f}$. Of course, $\theta_{\mathfrak{f}}^{*}(1)=0$.

Case of $\mathfrak{f}=210$. Here is our function:

$$\theta_{210}^{*}(1/u)=\begin{cases}1 & \text{if } 0<u\leq 1\\ 54/35 & \text{if } 1<u\leq 3\\ 57/35 & \text{if } 3<u\leq 7\\ 76/35 & \text{if } 7<u\leq 9\\ 79/35 & \text{if } 9<u\leq 79/8\\ 8u/35 & \text{if } 79/8\leq u\leq 10\end{cases}\qquad\begin{cases}16/7 & \text{if } 10<u\leq 13\\ 82/35 & \text{if } 13<u\leq 17\\ 94/35 & \text{if } 17<u\leq 41/2\\ 8u/35-2 & \text{if } 41/2\leq u\leq 22\\ 106/35 & \text{if } 22<u\leq 210\end{cases}$$

Polynomial approximation to $\theta_{\mathfrak{f}}^{*}(v)$. Starting from a polynomial approximation to $\theta_{\mathfrak{f}}^{*}(v)$ of the form

$$\left|\theta_{\mathfrak{f}}^{*}(v)-\sum_{0\leq r\leq R}\tilde{b}_{r}v^{r}\right|\leq\epsilon$$

for $0\leq v\leq V$ we infer the upper bound

$$(8.4)\qquad\qquad\theta_{\mathfrak{f}}^{*}(v)\leq\sum_{0\leq r\leq R}b_{r}v^{r}.$$

We build our approximation from Bernstein polynomials, since they are usually good candidates for approximating a continuous function in the L^{∞} sense. We let

$$(8.5)\qquad\qquad B_{n,k}(x)=\binom{n}{k}x^{k}(1-x)^{n-k}$$

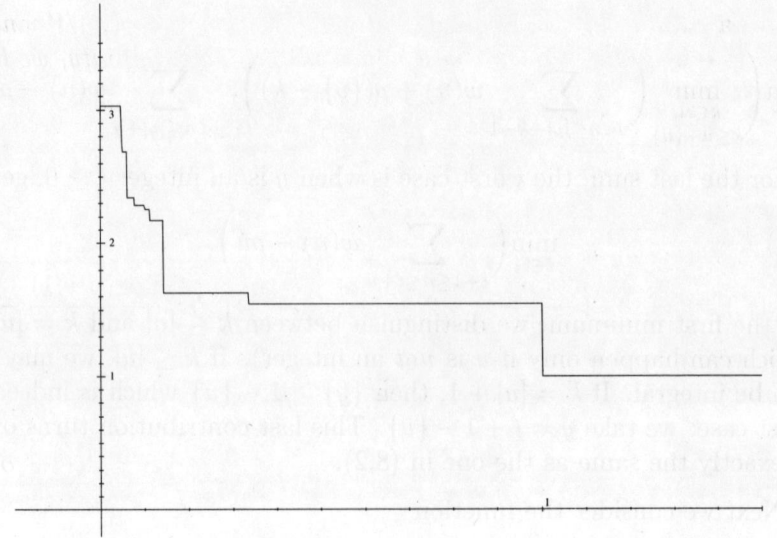

Figure 8.1. Graph of θ_{210}^*

and we consider

$$(8.6) \qquad B_n^* = \sum_{0 \le k \le n} B_{n,k}(v/V)\theta_{210}^*(Vk/n).$$

in order to approximate θ_{210}^* on $[0, V]$, where we shall choose V later on. But because of the discontinuities, this approximation cannot be closer than half the maximal jump, that is to say $\frac{1}{2}(76/35 - 57/35) = 19/70 = 0.27\ldots$. We can recover a part of this loss since we are only concerned with an upper bound of a nonincreasing function.

8.3. Some auxiliary estimates on multiplicative functions

We shall require some cumbersome estimates for certain multiplicative functions, and we prefer clearing these questions before entering into the main part of the proof. To alleviate somewhat the typographical work, we define

$$(8.7) \qquad \eta_r(k) = \prod_{p|k} \frac{1 + p^{r+1}}{p - 1} \quad , \quad \eta_r^b(k) = \prod_{p|k} \frac{1 + p^{r+2-2p^{r+1}}}{(p - 1)^2}.$$

Lemma 8.2. *Let* $\mathfrak{f}^*$ *be a positive integer. We set* $\rho^* = \phi(\mathfrak{f}^*)/\mathfrak{f}^*$ *and use* $t(q) = 1 - \sigma(q)/S^*$. *For any real number* S^* *going to infinity, we have*

$$\sum_{\substack{q/\sigma(q)\leq S^*, \\ (q,\mathfrak{f}^*)=1}} \frac{t(q)^2}{\phi(q)} = \rho^*(\mathrm{Log}\, S^* + \kappa(\mathfrak{f}^*)) + o(1)$$

with

$$\kappa(\mathfrak{f}^*) = \gamma + \sum_{p\geq 2}\left(\frac{\mathrm{Log}\, p}{p(p-1)} - \frac{\mathrm{Log}(1+p^{-1})}{p}\right) + \sum_{p|\mathfrak{f}^*}\frac{\mathrm{Log}(p+1)}{p} - \frac{3}{2}$$

$(\kappa(210) = 1.115\,37\ldots)$ *and*

$$\sum_{\substack{q/\sigma(q)\leq S^*, \\ (q,\mathfrak{f}^*)=1}} \eta_r(q)t(q) = \frac{\rho^*}{2(r+1)}\prod_{p\nmid\mathfrak{f}^*}\left(1 - \frac{p^r-1}{p^{r+1}(p+1)}\right)S^{*(r+1)}(1 + o(1)).$$

Proof. The first estimate comes from (Selberg, 1991), who follows a method already used by (Bateman, 1972). We follow closely Selberg's proof and get

$$\sum_{\substack{q/\sigma(q)\leq S^*, \\ (q,\mathfrak{f}^*)=1}} \frac{\eta_r(q)t(q)}{q^r} = \frac{\rho}{2}\prod_{p\nmid\mathfrak{f}^*}\left(1 - \frac{p^r-1}{p^{r+1}(p+1)}\right)S^* + o(S^*).$$

We conclude by using an integration by parts.   ◇◇◇

Note that the quantities we end up computing are the same as the ones that appear in (Selberg, 1991), though we have one less to handle.

We define

$$(8.8)\qquad C_r(\Delta) = \frac{\phi(\Delta)^2}{\Delta^2}\sum_{\delta_1\delta_2\delta_3|\Delta}\frac{\eta_r^b(\delta_1)\eta_r(\delta_2\delta_3)}{\sigma(\delta_1)^{2r+2}\sigma(\delta_2)^{r+1}\sigma(\delta_3)^{r+1}}$$

as well as

$$(8.9)\quad c_r(p) = (p-1)^2(p+1)^{2r+2}\big((p+1)^{r+1}(p-1) + 2p^{r+1} + 2\big)$$
$$+ \big(1 + p^{r+1}(p-2)\big)\big((p+1)^{r+1}(p-1) + p^{r+1} + 1\big)$$

The next lemma gives a multiplicative expression for C_r.

Lemma 8.3.

$$C_r(\Delta) = \prod_{p|\Delta}\frac{c_r(p)}{(p-1)p^2(p+1)^{3r+3}}$$

Proof. We start with δ_3:

$$\sum_{\delta_3|\Delta/(\delta_1\delta_2)} \frac{\eta_r(\delta_3)}{\sigma(\delta_3)^{r+1}} = \prod_{p|\Delta/(\delta_1\delta_2)} \left(1 + \frac{1+p^{r+1}}{(p+1)^{r+1}(p-1)}\right)$$

$$= \prod_{p|\Delta/(\delta_1\delta_2)} \frac{(p+1)^{r+1}(p-1)+p^{r+1}+1}{(p+1)^{r+1}(p-1)}.$$

Our sum reduces to

$$\prod_{p|\Delta} \frac{\big((p+1)^{r+1}(p-1)+p^{r+1}+1\big)(p-1)}{p^2(p+1)^{r+1}}$$

$$\times \sum_{\delta_1\delta_2|\Delta} \frac{\prod_{p|\delta_1}(1+p^{r+1}(p-2))\eta_r(\delta_2)}{\phi(\delta_1)^2\sigma(\delta_1)^{2r+2}\sigma(\delta_2)^{r+1}} \prod_{p|\delta_1\delta_2} \frac{(p+1)^{r+1}(p-1)}{(p+1)^{r+1}(p-1)+p^{r+1}+1}.$$

We continue with δ_2:

$$\sum_{\delta_2|\Delta/\delta_1} \frac{1+p^{r+1}}{(p-1)(p+1)^{r+1}} \frac{(p+1)^{r+1}(p-1)}{(p+1)^{r+1}(p-1)+p^{r+1}+1}$$

$$= \prod_{p|\Delta/\delta_1} \left(1 + \frac{p^{r+1}+1}{(p+1)^{r+1}(p-1)+p^{r+1}+1}\right)$$

$$= \prod_{p|\Delta/\delta_1} \frac{(p+1)^{r+1}(p-1)+2p^{r+1}+2}{(p+1)^{r+1}(p-1)+p^{r+1}+1}.$$

Hence $C_r(\Delta)$ reduces to

$$\prod_{p|\Delta} \frac{\big((p+1)^{r+1}(p-1)+2p^{r+1}+2\big)(p-1)}{p^2(p+1)^{r+1}}$$

$$\times \sum_{\delta_1|\Delta} \prod_{p|\delta_1} \frac{1+p^{r+1}(p-2)}{(p-1)^2(p+1)^{2r+2}} \frac{(p+1)^{r+1}(p-1)+p^{r+1}+1}{(p+1)^{r+1}(p-1)+2p^{r+1}+2}$$

which reads

$$\prod_{p|\Delta} \frac{\big((p+1)^{r+1}(p-1)+2p^{r+1}+2\big)(p-1)}{p^2(p+1)^{r+1}}$$

$$\times \frac{c_r(p)}{(p-1)^2(p+1)^{2r+2}\big((p+1)^{r+1}(p-1)+2p^{r+1}+2\big)}$$

$$= \prod_{p|\Delta} \frac{c_r(p)}{(p-1)p^2(p+1)^{3r+3}}$$

$\diamond\diamond\diamond$

8.4. Local models for the sequence of primes

8.4.1. Choice of the local system. First, some remarks on what "sieving" means. Sieving is about gaining information on a sequence from what we know of it modulo d for several d's. If one looks at the sequence of primes modulo d and if we neglect the prime divisors of d, it simply is the set of reduced residue classes modulo d, which we have called $\mathcal{U}_d$. Thus, on the one hand we have the characteristic function of primes in the interval $[M + 1, M + N]$, say f, and on the other hand the characteristic function φ_d of the integers in this interval that are coprime to d for all $d \leq \sqrt{N}$. Notice here that it is enough to restrict our attention to squarefree d's.

Recalling what we did in section 1.1, we could simply try to get an approximation to f in terms of the φ_d's. However, the study there is patterned for almost orthogonal φ_q's, which is not the case of the sequence $(\varphi_d)_d$: if $q|d$, knowing that a given integer is coprime with d implies it is coprime with q, so there is redundancy of information. It implies in turn that these functions are far from being linearly independent. We unscrew the situation in the following way. When d is squarefree, we set

$$(8.10) \qquad \frac{d}{\phi(d)}\varphi_d = \sum_{q|d} \varphi_q^*$$

where

$$(8.11) \qquad \varphi_q^*(n) = \mu(q)c_q(n)/\phi(q)$$

and $c_q(n)$ is Ramanujan sum given by

$$(8.12) \qquad c_q(n) = \sum_{a \bmod^* q} e(na/q) = \sum_{\ell|q} \ell\mu(q/\ell).$$

Verifying (8.10) is easy:

$$\sum_{q|d} \mu(q)c_q(n)/\phi(q) = \prod_{p|d}\left(1 - \begin{cases} 1 & \text{if } p|n \\ -1/(p-1) & \text{otherwise} \end{cases}\right).$$

To understand better the functions φ_q^* defined by (8.10), the reader may consult section 11.8 and in particular equation (11.33).

Here is our set of moduli q:

$$(8.13) \qquad \{q \mathbin{/} \sigma(q) \leq S, \mu^2(q) = 1, (q, \mathfrak{f}) = 1\},$$

where $\sigma(q) = \sum_{d|q} d$. The reason for this choice will become clear later on.

8.4.2. Study of the local models. Note that

$$(8.14) \qquad [\varphi_q^* | \varphi_{q'}^*] = \frac{\mu(q)}{\phi(q)} \frac{\mu(q')}{\phi(q')} \sum_n w(n) c_q(n) c_{q'}(n).$$

We note that when q and q' have a prime factor in common, say δ, then $c_\delta(n)^2 = \phi((n,\delta))^2$ would factor out: this contribution is non-negative and we use this fact here. Let Δ be a squarefree integer coprime with $\mathfrak{f}$. Write $(q, q', \Delta) = \delta$, so that $[\varphi_q^* | \varphi_{q'}^*]$ equals

$$(8.15)$$
$$\frac{\mu(q)\mu(q')}{\phi(q)\phi(q')} \sum_{\substack{\ell | q/\delta \\ \ell' | q'/\delta \\ h | \delta}} \ell\mu\left(\frac{q}{\ell}\right)\ell'\mu\left(\frac{q}{\ell'}\right)(\mu \star \phi^2)(h)\left(\frac{\rho N}{h[\ell,\ell']} + R_{h[\ell,\ell']}(M,N,\mathfrak{f})\right)$$

where

$$(8.16) \qquad R_d(M,N,\mathfrak{f}) = \sum_{\substack{M+1 \le n \le M+N \\ d | n}} w(n) - \frac{\rho N}{d}.$$

The reader will check that the main term (corresponding to $\rho N/[\ell,\ell']$) vanishes if $q \ne q'$ and is $\rho N/\phi(q)$ otherwise. We carry over this change to the bilinear form $\left\| \sum_q \xi_q \varphi_q^* \right\|^2$, which equals to the diagonal term $\rho N \sum_q |\xi_q|^2 / \phi(q)$ to which we add

$$\mathfrak{R} = \sum_{\delta_1\delta_2\delta_3 | \Delta} \frac{\mu(\delta_2\delta_3)}{\phi(\delta_1)^2\phi(\delta_2)\phi(\delta_3)} \sum_{\substack{(\ell, \mathfrak{f}\Delta)=1 \\ (\ell', \mathfrak{f}\Delta)=1}} \frac{\mu(\ell)\xi_{\delta_1\delta_2\ell}}{\phi(\ell)} \frac{\mu(\ell')\xi_{\delta_1\delta_3\ell'}}{\phi(\ell')}$$

$$\times \sum_{\substack{d | \ell\delta_2 \\ d' | \ell'\delta_3 \\ h | \delta_1}} dd'\mu(\ell\delta_2/d)\mu(\ell'\delta_3/d')(\mu \star \phi^2)(h) R_{h[d,d']}(M,N,\mathfrak{f}).$$

The simplicity of the method is somewhat obscured by the precise handling of $\mathfrak{R}$, but this is the price we pay for an improved bound. However the reader might want to start with $\Delta = 1$ and bound $R_d(M,N,\mathfrak{f})$ by $\mathcal{O}(1)$. We may even use what follows in this special case: simply take $R = 1$ and $b_0 = 2$ in (8.17). In the general case, we treat the error term by invoking say (8.4):

$$(8.17) \qquad \left| R_{h[d,d']}(M,N,\mathfrak{f}) \right| \le \theta_{\mathfrak{f}}^*(h[d,d']/N) \le \sum_{0 \le r \le R} b_r (h[d,d']/N)^r.$$

We infer

$$\mathfrak{R} \leq \sum_{0 \leq r \leq R} b_r N^{-r} \sum_{\delta_1 \delta_2 \delta_3 | \Delta} \frac{1}{\phi(\delta_1)^2 \phi(\delta_2) \phi(\delta_3)} \sum_{\substack{(\ell, \mathfrak{f}\Delta)=1 \\ (\ell', \mathfrak{f}\Delta)=1}} \frac{|\xi_{\delta_1 \delta_2 \ell}|}{\phi(\ell)} \frac{|\xi_{\delta_1 \delta_3 \ell'}|}{\phi(\ell')}$$

$$\times \sum_{\substack{d | \ell \delta_2 \\ d' | \ell' \delta_3 \\ h | \delta_1}} dd' (\mu \star \phi^2)(h) h^r [d, d']^r.$$

Recalling (8.7), it is straightforward to simplify the coefficient of $b_r N^{-r}$ into

$$\sum_{\delta_1 \delta_2 \delta_3 | \Delta} \eta_r^\flat(\delta_1) \eta_r(\delta_2 \delta_3) \sum_{\substack{(\ell, \mathfrak{f}\Delta)=1 \\ (\ell', \mathfrak{f}\Delta)=1}} |\xi_{\delta_1 \delta_2 \ell}| \eta_r(\ell) |\xi_{\delta_1 \delta_3 \ell'}| \eta_r(\ell')$$

$$\times \prod_{p | (\ell, \ell')} \frac{1 + 2p^{r+1} + p^{r+2}}{(1 + p^{r+1})^2}.$$

The factor that depends on (ℓ, ℓ') is somewhat troublesome. We handle it in the following way: for $r = 0$, it is equal to 1. Otherwise, let P be the smallest prime number that does not divide $\mathfrak{f}\Delta$. This prime factor is going to go to infinity, and we approximate the factor depending on (ℓ, ℓ') essentially by $1 + \mathcal{O}(P^{-1})$. More precisely, we write

$$\sum_{(\ell \ell', \mathfrak{f}\Delta)=1} |\xi_{\delta_1 \delta_2 \ell}| \eta_r(\ell) |\xi_{\delta_1 \delta_3 \ell'}| \eta_r(\ell') \left| \prod_{p | (\ell, \ell')} \frac{1 + 2p^{r+1} + p^{r+2}}{(1 + p^{r+1})^2} - 1 \right|$$

$$\ll_r \sum_{p \geq P} \sum_{\substack{(m, p\mathfrak{f}\Delta)=1, \\ (m', p\mathfrak{f}\Delta)=1}} |\xi_{\delta_1 \delta_2 pm}| \eta_r(pm) |\xi_{\delta_1 \delta_3 pm'}| \eta_r(pm')$$

$$\ll_r \sum_{p \geq P} p^{2r} \sum_{m, m'} |\xi_{\delta_1 \delta_2 pm}| \eta_r(m) |\xi_{\delta_1 \delta_3 pm'}| \eta_r(m')$$

The idea here is that the factor $\xi_{\delta_1 \delta_2 pm}$ forces m to be rather small. Indeed, anticipating the values of ξ in (8.18) and using Lemma 8.2, we get the above to be not more than

$$\left(\frac{Z}{\rho N} \right)^2 \sum_{p \geq P} p^{2r} (S/p)^{2r+2} \ll_r \left(\frac{Z}{\rho N} \right)^2 S^{2r+2} P^{-1}.$$

This will give rise to the error term

$$\left(\frac{Z}{\rho N} \right)^2 \sum_{\delta_1 \delta_2 \delta_3 | \Delta} \frac{\eta_r^\flat(\delta_1) \eta_r(\delta_2 \delta_3)}{\sigma(\delta_1 \delta_2)^{r+1} \sigma(\delta_1 \delta_3)^{r+1}} \sum_{1 \leq r \leq R} \frac{S^{2r+2} |b_r|}{N^r P}$$

which up to a multiplicative constant is not more than

$$\left(\frac{Z}{\rho N}\right)^2 \prod_{p|\Delta}(1+p^{-1})^2 \sum_{1\leq r\leq R} \frac{S^{2r+2}|b_r|}{N^r P}.$$

The factor P^{-1} will indeed be enough to control this quantity. Hence, again anticipating (8.18), we reach

$$\left\|\sum_q \xi_q \varphi_q^*\right\|^2 \leq \rho N \sum_q |\xi_q|^2/\phi(q)$$

$$+ \sum_{0\leq r\leq R} \frac{b_r}{N^r} \sum_{\delta_1\delta_2\delta_3|\Delta} \eta_r^b(\delta_1) \sum_{\substack{(\ell,\mathfrak{f}\Delta)=1,\\(\ell',\mathfrak{f}\Delta)=1}} |\xi_{\delta_1\delta_2\ell}|\eta_r(\delta_2\ell)|\xi_{\delta_1\delta_3\ell'}|\eta_r(\delta_3\ell')$$

$$+ \mathcal{O}\left(\left(\frac{Z}{\rho N}\right)^2 \prod_{p|\Delta}(1+p^{-1})^2 \sum_{1\leq r\leq R} \frac{S^{2r+2}|b_r|}{N^r P}\right).$$

8.5. Using the hermitian inequality

Optimizing in ξ is too difficult. We stick to the simplest choice: $M_i = \rho N/\phi(q)$, $[f|\varphi_i^*]/M_i = Z/(\rho N)$, $n_i = \sigma(q)/\phi(q)$ and $Y = Z/S$.

(8.18)
$$\xi_q = \frac{Z}{\rho N}t(q), \quad t(q) = 1 - \frac{\sigma(q)}{S}$$

for a parameter S we shall choose later on.

We invoke Lemma 8.2 to compute $\sum_{(\ell,\mathfrak{f}\Delta)=1} |\xi_{\delta_1\delta_2\ell}|\eta_r(\ell)$ with $S^* = S/\sigma(\delta_1\delta_2)$ and $\mathfrak{f}^* = \mathfrak{f}\Delta$. There appear constants in the form of an Euler product, say $\mathfrak{S}_r(\mathfrak{f}^*)$, which we again approximate by $1 + \mathcal{O}(P^{-1})$. In a first step we reach

$$Z \geq \left(\frac{Z}{\rho N}\right)^2 \rho^2 N \left(\text{Log } S + \kappa(\mathfrak{f})\right) + \frac{2Z^2}{\rho N S} \sum_{(q,\mathfrak{f})=1} \frac{\sigma(q)t(q)}{\phi(q)}$$

$$- \sum_{0\leq r\leq R} \frac{Z^2 b_r}{\rho^2 N^{r+2}} \sum_{\delta_1\delta_2\delta_3|\Delta} \eta_r^b(\delta_1) \sum_{\substack{(\ell,\mathfrak{f}\Delta)=1\\(\ell',\mathfrak{f}\Delta)=1}} t(\delta_1\delta_2\ell)\eta_r(\ell\delta_2)t(\delta_1\delta_3\ell')\eta_r(\ell'\delta_3)$$

$$+ \mathcal{O}\left(\left(\frac{Z}{\rho N}\right)^2 \prod_{p|\Delta}(1+p^{-1})^2 \sum_{1\leq r\leq R} \frac{S^{2r+2}|b_r|}{N^r P}\right).$$

After some rearrangement, we obtain:

$$N/Z \geq \operatorname{Log} S + \kappa(\mathfrak{f}) + 1 - \sum_{0 \leq r \leq R} \frac{b_r (S^2/N)^{r+1}}{4(r+1)^2} C_r(\Delta)^2 \mathfrak{S}_r(\mathfrak{f}\Delta)^2$$

$$+ \mathcal{O}\left(\prod_{p|\Delta}(1 + p^{-1})^2 P^{-1} \sum_{1 \leq r \leq R} |b_r|(S^2/N)^{r+1} \right) + o(1)$$

And since $\mathfrak{S}_r(\mathfrak{f}\Delta) = 1 + \mathcal{O}(P^{-1})$, we finally reach

$$N/Z - \tfrac{1}{2}\operatorname{Log} N \geq \tfrac{1}{2}\operatorname{Log}(S^2/N) + \kappa(\mathfrak{f}) + 1 - \sum_{0 \leq r \leq R} \frac{b_r(S^2/N)^{r+1}}{4(r+1)^2} C_r(\infty/\mathfrak{f})^2$$

$$+ \mathcal{O}\left(\prod_{p|\Delta}(1+p^{-1})^2 P^{-1} \sum_{1 \leq r \leq R} |b_r|(S^2/N)^{r+1} \right) + o(1).$$

At this level, we send Δ (and P) to infinity and we are left with finding an optimal value for S^2/N. It would be satisfactory to have an expression for the final constant, but we are not able to reach such precision. In particular, the b_r's should not appear in such an expression. We are, however, able to get numerical results.

Some numerical results:

n	V	S^2/N	
10	1.2	0.883 867	2.958 900
40	1.2	0.903 740	2.990 585
60	1.2	0.922 038	3.004 986
100	1.2	0.923 831	3.009 657
100	1.1	0.926 587	3.010 536

8.6. Generalization to a weighted sieve bound

We anticipate somehow the forthcoming chapters. To get similar results in the general case, we would start from (11.21) with ψ_q^* defined in (11.13). When sieving an interval, $|R([\ell, \ell'])|$ can be bounded by $|\mathcal{L}_\ell||\mathcal{L}_{\ell'}|$, and some work later, we end up in the situation of a mixed almost orthogonal system as in section 1.1. Following the theory therein, we end up with a weighted sieve bound as in the example above. We should add that (Montgomery & Vaughan, 1973) (see also (Preissmann, 1984)) already gave weighted bounds, and for instance, (Siebert, 1976) employed them to prove a neat upper bound for the number of twin primes, see section 21.3. Note that these weights do not depend on the used compact set. The path presented here is incomplete in more than one aspect, and the main deficiency being that fairly intricate averages are required, similar to the ones studied in Lemma 8.2, nevertheless, it *does* lead to a weighted bound depending on $\mathcal{K}$.

9 Twin primes and local models

We saw in the previous section, and in an extremely simple example, how local models enter into the game of sieving. Further, we took the opportunity of exploring somewhat more intricate weights. While doing this, we missed one crucial fact: the good almost orthogonality bounds for our local models in the previous chapter come from the simple structure of the set we are sieving, as will be more evident in Lemma 19.4. Technically speaking, the expression for c_q in terms of additive characters has $\phi(q)$ summands, while the one in terms of divisors (8.12) has only $2^{\omega(q)}$ summands. We now give further details in the case of prime twins, where this feature will clearly show up. A general treatment is given in section 11.6.

We prove here the following classical result:

Theorem 9.1. *The number of primes p in the interval $[M, M+N]$ that are such that $p+2$ is also a prime number is not more than*

$$(16 + o(1)) \prod_{p \geq 3} \left(1 - \frac{1}{(p-1)^2}\right) \frac{N}{\mathrm{Log}^2 N}$$

where the $o(1)$ denotes a quantity that goes to 0 when N goes to infinity.

This bound is believed to be 8 times too large. The case $M = 0$ has seen a number of refinements: using the Bombieri-Vinogradov Theorem directly reduces this bound by 2 (case $M = 0$) and further works led to reduce the $16 + o(1)$, among which we select the reduction to $7.835 + o(1)$ due to (Chen, 1978), to 6.836 due to (Wu, 1990), recently to $6.812 + o(1)$ by (Cai & Lu, 2003) and even more recently to $6.7992 + o(1)$ by (Wu, 2004).

9.1. The local model for twin primes

Let us first define our set of moduli:

(9.1) $$\mathcal{Q} = \{q \leq Q, \; q \text{ odd and squarefree}\}.$$

To each couple (q, d) where q in $\mathcal{Q}$ and d is a divisor of q, we associate $u_{q,d}$ the unique integer between 1 and d such that $(q/d)u_{q,d} \equiv 1[d]$. Our local model is then

(9.2)
$$\varphi_q^*(n) = \frac{\mu(q)}{\phi_2(q)} \sum_{d|q} c_q(n + 2u_{q,d}q/d)$$

with $\phi_2(q) = \prod_{p|q}(p-2)$. We take the simplest hermitian product, namely

$$[f|g] = \sum_{\substack{M < n \le M+N, \\ (n,2)=1}} f(n)\overline{g(n)}.$$

The next step is to evaluate pairwise the scalar products of our local models:

$$[\varphi_q^*|\varphi_{q'}^*] = \frac{\mu(q)\mu(q')}{\phi_2(q)\phi_2(q')} \sum_{\substack{d|q, \\ d'|q'}} \sum_n c_q(n + 2u_{q,d}q/d)c_{q'}(n + 2u_{q',d'}q'/d')$$

$$= \frac{\mu(q)\mu(q')}{\phi_2(q)\phi_2(q')} \sum_{\substack{d|q, \ \delta|q, \\ d'|q' \ \delta'|q'}} \delta\mu(q/\delta)\delta'\mu(q'/\delta') \sum_{n / \left\{ \begin{array}{l} n \equiv -2u_{q,d}q/d[\delta], \\ n \equiv -2u_{q',d'}q'/d'[\delta'] \end{array} \right.} 1.$$

The last two congruences are not always compatible: if p divides δ and δ', and if it divides q/d and q'/d', both congruences reduce to $n \equiv 0[p]$. If p divides neither q/d nor q'/d', then the congruences reduce to $n \equiv -2[p]$. Which means we need p to divide $(q/d, q'/d')$ or (d, d'). As a result, we infer that $[\varphi_q^*|\varphi_{q'}^*]$ equals

$$\frac{N\mu(q)\mu(q')}{\phi_2(q)\phi_2(q')} \sum_{\substack{d|q, \\ d'|q' \ (\delta,\delta')|(d,d')(q/d,q'/d')}} \sum_{\delta|q,\delta'|q'} \frac{\delta\mu(q/\delta)\delta'\mu(q'/\delta')}{[\delta,\delta']}$$

$$+ \mathcal{O}^*\left(\frac{2^{\omega(q)}\sigma(q)}{\phi_2(q)} \frac{2^{\omega(q')}\sigma(q')}{\phi_2(q')} \right).$$

We evaluate the arithmetic part of the main term by plugging the summations over d and d' inside: the part of d that divides $q/(\delta,\delta')$ is freely chosen, giving $2^{\omega(q)-\omega((\delta,\delta'))}$ choices, and similarly for d' with q'. Next a prime divisor of δ and δ' either divides both of d and d' or divides none of them. Thus there is a divisor, say h, of (δ,δ') that divides exactly d and d'. We have $2^{\omega((\delta,\delta'))}$ such divisors. Collecting these observations, we readily discover our inner sum to be equal to $2^{\omega(q)+\omega(q')-\omega((\delta,\delta'))}$ so that we get

$$[\varphi_q^*|\varphi_{q'}^*] = \frac{2^{\omega(q)+\omega(q')}N}{\phi_2(q)\phi_2(q')} \sum_{\delta|q,\delta'|q'} \frac{\delta\mu(\delta)\delta'\mu(\delta')}{[\delta,\delta']2^{\omega((\delta,\delta'))}} + \mathcal{O}^*\left(\frac{2^{\omega(q)}\sigma(q)}{\phi_2(q)} \frac{2^{\omega(q')}\sigma(q')}{\phi_2(q')} \right).$$

When there is a prime that divides q but not q', the main term vanishes. We are thus left with the case $q = q'$, getting

$$(9.3) \qquad [\varphi_q^*|\varphi_{q'}^*] = \frac{2^{\omega(q)} N \mathbb{1}_{q=q'}}{\phi_2(q)} + \mathcal{O}^*\left(\frac{2^{\omega(q)}\sigma(q)}{\phi_2(q)}\frac{2^{\omega(q')}\sigma(q')}{\phi_2(q')}\right).$$

Concerning the almost orthogonality hypothesis, we take the easiest way out: we set $M_q = 2^{\omega(q)} N/\phi_2(q)$ and send the error term into the bilinear form, i.e. we write

$$(9.4) \qquad \left|\sum_q \xi_q \varphi_q^*\right|^2 \le \sum_q M_q|\xi_q|^2 + \sum_{q,q'} \xi_q \overline{\xi_{q'}} m_{q,q'}$$

with

$$(9.5) \qquad |m_{q,q'}| \le \frac{2^{\omega(q)}\sigma(q)}{\phi_2(q)}\frac{2^{\omega(q')}\sigma(q')}{\phi_2(q')}.$$

9.2. Estimation of the remainder term

To handle the error term, we are to compute or at least give an upper bound for the average

$$\sum_{q \in \mathcal{Q}} \mu^2(q) 2^{\omega(q)}\sigma(q)/\phi_2(q).$$

This is standard theory: one possibility would be to first evaluate the average of the summand above divided by q via the convolution method as in section 5.3 and then recover the one we are interested in by a summation by parts. The Levin-Fainleib like theorem presented in chapter 21 as Theorem 21.1 would also suffice: however the summation by parts would lead to a cancellation of the "main terms", leaving us only with a $\mathcal{O}$-result of the good order of magnitude, while Theorem 21.2 or the convolution method would give rise to an asymptotic expression. We present an alternative path that also leads only to an upper bound. First, we prove the following theorem that relies on a theme initially developed in (Hall, 1974). The best result in this direction is in (Halberstam & Richert, 1979). Of course, we also extend it to encompass values at powers of primes. The starting idea is still taken from the celebrated (Levin & Fainleib, 1967).

Theorem 9.2. *Let $D \ge 2$ be a real parameter. Assume g is a multiplicative non-negative function such that*

$$\sum_{\substack{p\ge 2, \nu\ge 1 \\ p^\nu \le Q}} g(p^\nu) \operatorname{Log}(p^\nu) \le KQ + K' \qquad (\forall Q \in [1, D])$$

for some constants $K, K' \geq 0$. Then for $D > \exp(K' - 1)$, we have

$$\sum_{d \leq D} g(d) \leq \frac{(K+1)D}{\operatorname{Log} D - K' + 1} \sum_{d \leq D} g(d)/d.$$

Proof. Let us set $\tilde{G}(D) = \sum_{d \leq D} g(d)/d$. Using $\operatorname{Log} \frac{D}{d} \leq \frac{D}{d} - 1$, we get

$$G(D) \operatorname{Log} D = \sum_{d \leq D} g(d) \operatorname{Log} \frac{D}{d} + \sum_{d \leq D} g(d) \operatorname{Log} d$$

$$\leq D\tilde{G}(D) - G(D) + \sum_{\substack{p \geq 2, \nu \geq 1 \\ p^\nu \leq D}} g(p^\nu) \operatorname{Log}(p^\nu) \sum_{\substack{\ell \leq D/p^\nu \\ (\ell, p) = 1}} g(\ell)$$

where we get the second summand by writing $\operatorname{Log} d = \sum_{p^\nu \| d} \operatorname{Log}(p^\nu)$. Finally

$$\sum_{\substack{p \geq 2, \nu \geq 1 \\ p^\nu \leq D}} g(p^\nu) \operatorname{Log}(p^\nu) \sum_{\substack{\ell \leq D/p^\nu \\ (\ell, p) = 1}} g(\ell) = \sum_{\ell \leq D} g(\ell) \sum_{\substack{p \geq 2, \nu \geq 1 \\ p^\nu \leq D/\ell \\ (p, \ell) = 1}} g(p^\nu) \operatorname{Log}(p^\nu)$$

$$\leq \sum_{\ell \leq D} g(\ell) \left(\frac{KD}{\ell} + K' \right)$$

from which the theorem follows readily. $\diamond\diamond\diamond$

Once we apply this result, we are again left with getting an upper bound for the average of $\mu^2(q) 2^{\omega(q)} \sigma(q)/(q\phi_2(q))$, where we apply Theorem 21.1. As a result, we get the bound

(9.6) $$\sum_{q \in \mathcal{Q}} \mu^2(q) 2^{\omega(q)} \sigma(q)/\phi_2(q) \ll Q \operatorname{Log}(3Q).$$

9.3. Main proof

Let f be the characteristic function of the twin primes in our interval. Note that

(9.7) $$[f | \varphi_q^*] = \mu^2(q) 2^{\omega(q)} Z/\phi_2(q)$$

with $Z = \sum_n f(n)$. We apply Lemma 1.2. We have $\xi_q = \mu^2(q) Z/N$, so we get

$$G_1(Q) Z^2/N \leq Z + (Z/N)^2 \left(\sum_q 2^{\omega(q)} \sigma(q)/\phi_2(q) \right)^2$$

$$\leq Z + \mathcal{O}((ZN^{-1} Q \operatorname{Log} Q)^2)$$

with

(9.8) $$G_1(Q) = \sum_{q \in \mathcal{Q}} 2^{\omega(q)}/\phi_2(q)$$

which is evaluated by standard means. We give such an evaluation in chapter 21. We even consider this very special case in section 21.3 where we show how the standard sieve bound (that is Corollary 2.1, page 22) works on this special case. The evaluation of the remainder term comes from (9.6), and we reach

$$\left(G_1(Q) - \mathcal{O}(N^{-1}Q^2 \operatorname{Log}^2 Q)\right)Z \leq N$$

with

$$G_1(Q) \sim \frac{1}{4} \prod_{p \geq 3} \frac{(p-1)^2}{p(p-2)} \operatorname{Log}^2 Q.$$

We can thus take $Q = o(\sqrt{N})$, improving on the usual proof via Selberg's sieve which a priori only allows for $Q = o(\sqrt{N}/\operatorname{Log} N)$. When carefully studied, a similar improvement is accessible there, as shown in the notes of the corresponding chapter of (Halberstam & Richert, 1974).

9.4. Guessing the local model

We simply exhibited φ_q^*, taken straight from our hat... But now the proof has been shown to function, some more explanations are surely called for! As a matter of fact, most of the mystery gets cleared once we rewrite $c_q(n + 2u_{q,d}q/d)$ in a multiplicative form (remember that since q is squarefree, d and q/d have distinct prime factors):

$$c_q(n + 2u_{q,d}q/d) = \prod_{\substack{p|d \\ p|n+2}} (p-1) \prod_{\substack{p|d \\ p\nmid n+2}} (-1) \prod_{\substack{p|q/d \\ p|n}} (p-1) \prod_{\substack{p|q/d \\ p\nmid n}} (-1)$$

$$= \mu(q) \prod_{\substack{p|d \\ p|n+2}} (1-p) \prod_{\substack{p|q/d \\ p|n}} (1-p)$$

so that

$$\varphi_q^*(n) = \frac{1}{\phi_2(q)} \prod_{p|q} \left(1 - p\mathbb{1}_{p|n+2} + 1 - p\mathbb{1}_{p|n}\right) = \frac{1}{\phi_2(q)} \prod_{p|q} \left(2 - p\mathbb{1}_{\mathcal{L}_p}(n)\right)$$

and this in turn implies (recall that $|\mathcal{K}_r| = \phi_2(r)$)

(9.9) $$\sum_{q|r} \varphi_q^*(n) = \prod_{p|r} \left(1 + \frac{2 - p\mathbb{1}_{\mathcal{L}_p}(n)}{p-2}\right) = \frac{r}{|\mathcal{K}_r|} \mathbb{1}_{\mathcal{K}_r}(n)$$

unveiling at once most of the hidden mechanism! The correcting coefficient $r/|\mathcal{K}_r|$ has however still to be explained: imagine we were starting from $\mathbb{1}_{\mathcal{K}_r}(n)$ and considered "the solution" $(\varphi_q^*)_{q|r}$ of (9.9). Then we would discover that each φ_q^* in fact depend on r. A way to explain the correcting coefficient is simply to say that, with it, this dependance disappears. But, why does it indeed disappear? One way of explaining this fact is to say that this function is invariant under the operators $J_{\tilde{d}}^{\tilde{q}}$ introduced in chapter 4 (with $\mathcal{K} = \hat{\mathbb{Z}}$), so that (9.9) is simply the Fourier decomposition. We can however go one step further and precisely point out where this invariance comes from: we are to have

$$[c(q)\mathbb{1}_q|L_{\tilde{q}}^{\tilde{d}}f]_q = [c(d)\mathbb{1}_d|f]_d$$

for $d|q$, some coefficients $(c(d))_{d|q}$ and every time for every function f that depends only on its argument modulo d. This equation reads

$$\frac{c(q)}{q} \sum_{a \in \mathcal{K}_q} f(a \bmod d) = \frac{c(d)}{d} \sum_{b \in \mathcal{K}_d} f(b).$$

On taking for f the characteristic function of a single point modulo d, we see that $c(q)/q = 1/|\mathcal{K}_q|$ is the only choice. As a by-product of this construction, we see that we can only extend the method to compacta that verify the Johnsen-Gallagher condition.

9.5. Prime k-tuples

We were looking at pairs $(n, n+2)$ for which each component is prime. Extending the problem to k-tuples means looking for infinitely many integers n for which all the components of $(n+h_1, \ldots, n+h_k)$ are simultaneously prime. Determining which tuples $(h_1, \ldots, h_k)$ should have this property is a non trivial problem; Notice first that $(0, 1)$ is clearly not a good choice! Here the obstruction comes from what happens modulo 2. In general the conjecture known as *the prime k-tuples conjecture*, first stated by (Hardy & Littlewood, 1922) is that obtructions can only be local. This warrants a definition:

Definition 9.1. *A k-tuple $(h_1, \ldots, h_k)$ of increasing integers is said to be a k-tuple of admissible shifts if the set $\{h_1, \ldots, h_k\}$ does not cover all of $\mathbb{Z}/p\mathbb{Z}$ for any prime p.*

The *length* of such tuple of a admissible shifts being $h_k - h_1 + 1$, it is enough to restrict p to be not more than this length in the statement. For example $(0, 2, 6, 9, 12)$ is admissible of length 13.

An interesting problem is to find as dense as possible such tuples, where the density is best quantified in terms of the length $N = h_k - h_1 + 1$ compared to the number of primes less than N, which we denote exceptionally here by $\pi(N)$. (Hensley & Richards, 1974) proved that there exist k-tuples of admissible shifts of size

$$k \geq \pi(N) + (\text{Log } 2 - \varepsilon)\frac{N}{\text{Log}^2 N}$$

for every $\varepsilon > 0$ and provided N is large enough in terms of ε. Note that the Brun-Titchmarsh Theorem says that k is bounded by $2N/\text{Log } N$. If one is ready to believe the prime k-tuple conjecture, such extreme examples of admissible shifts thus provides us with a lower bound for the best possible upper bound in the Brun-Titchmarsh Theorem. In order to avoid to appeal to the prime k-tuple conjecture, it would be necessary to indeed exhibit specific examples of such tuples, but this is still beyond the power of nowadays algorithms and computers. As of today, the best

(Dusart, 1998) has built a 1 415-uple of admissible shifts of length 11 763, while $\pi(11\,763) = 1\,409$, but no one has been yet able to produce a corresponding prime 1 415-uple. The reader will find on the site of (Forbes, n.d.) a list of long prime tuples, for instance:

1 906 230 835 046 648 293 290 043 $+ 0, 4, 6, 10, 16, 18, 24, 28,$

$30, 34, 40, 46, 48, 54, 58, 60, 66, 70$

due to J. Waldvogel & P. Leikauf in 2001. It contains 18 primes for a length of 70, while $\pi(70) = 19$.

Let us mention finally (Elsholtz, 2004) where the reader will find another application of sieve technique to k-tuple problems, but this time with a k of size $\text{Log } N$ for primes of size N.

10 The three primes theorem

We prove here the celebrated theorem of (Vinogradov, 1937):

Theorem 10.1.
Every large enough odd integer is a sum of three prime numbers.

The proof we present uses our large sieve setting intensively, both the large sieve inequality and the notion of local models. The novelty here is in dispensing with the circle method. Proofs exhibiting such a feature have already been given by both (Heath-Brown, 1985) and (Iwaniec, 1994) via, if not exactly, the dispersion method, or at least using ideas derived from it. The first author establishes as a preliminary step an estimate for the L^2-mean of the number of representations as a sum of two primes, as we do here, while the second one goes directly to the number of representations of an integer as a sum of three primes. We use yet another path, though part of the techniques developed here are inspired by chapter XX of (Iwaniec & Kowalski, 2004).

The proof will unfold in two steps: we first prove the required asymptotic for

$$(10.1) \qquad \mathfrak{R} = \sum_m r_2(m)^2 \quad \text{where} \quad r_2(m) = \sum_{n_1+n_2=m} \tilde{\Lambda}(n_1)\tilde{\Lambda}(n_2)$$

and $\tilde{\Lambda}(n) = \Lambda(n)F_N(n)$, F_N being a smoothing function described in next section. The forthcoming proof will show a use of the large sieve inequality close to that of the Parseval identity. It has already been partially used in (Ramaré, 1995). The asymptotic for $\mathfrak{R}$ also implies the following result.

Theorem 10.2 (Tchudakov, van der Corput, Estermann).
Almost every even integer is a sum of two prime numbers.

We sketch the proof in section 10.6. In the second step we shall prove the three primes theorem. We shall use a local model for the (suitably modified) number of representations of an integer as a sum of two primes and *not* for the primes. It turns out that both are proportional here, up to the infinite factor.

10.1. An approximate Bessel inequality

Let us keep the notations of Lemma 1.2 and consider the following hermitian products:

$$(10.2) \qquad \langle f|g\rangle = \sum_i M_i^{-1}[f|\varphi_i^*]\overline{[g|\varphi_i^*]}$$

and

$$(10.3) \qquad [\![f|g]\!] = [f|g] - \langle f|g\rangle + \sum_{i,j} \xi_i(f)\overline{\xi_j(g)}\omega_{i,j}.$$

Lemma 1.2 tells us that this last one is non-negative, so that we may apply the Cauchy-Schwarz inequality. When the contribution of the $\omega_{i,j}$'s is indeed an error term, and when $\langle f|f\rangle$ approaches $\|f\|_2^2$ "sufficiently well", then $[\![f|f]\!]$ is small and $\langle f|g\rangle$ is an approximation to $[f|g]$ for reasonable g's. This is the key to our approach to some binary additive problems.

10.2. Some Fourier analysis to handle the size condition

The function F we use is essentially Fejer kernel and its graph is the one below.

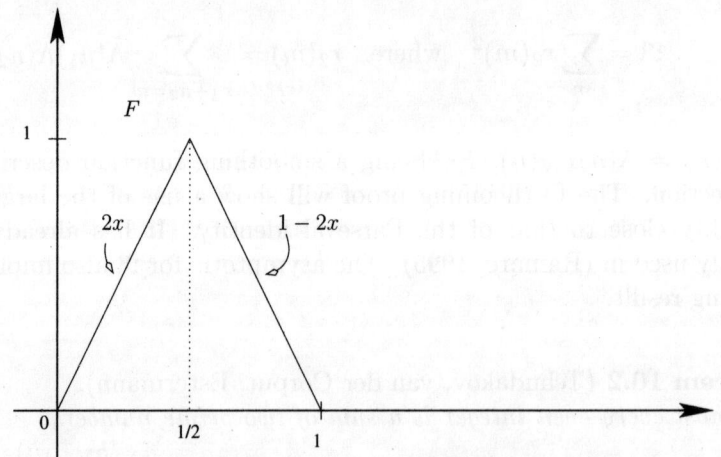

Its Fourier transform is given by

$$(10.4) \qquad \hat{F}(y) = \int_{-\infty}^{\infty} F(x)e(xy)dx = e(y/2)\left(\frac{\sin\frac{\pi y}{2}}{\pi y}\right)^2.$$

Fourier inversion yields

$$(10.5) \qquad F(x) = \int_{-\infty}^{\infty} \hat{F}(y) e(-xy) dy.$$

We also set $F_N(x) = F(x/N)$. Finally

$$(10.6) \qquad \int_{-\infty}^{\infty} \left(\frac{\sin \frac{\pi}{2} y}{\pi y} \right)^8 dy = \frac{151}{40\,320}$$

a constant we shall meet at several different places, and which we call C_0.

Of course, this function is nothing special and we could have chosen any function that vanishes at 0 and 1 whose Fourier transform decreases fast enough. This smoothing function is introduced to handle the size condition $0 \le n_1 \le N$ on all our variables. By approximating the characteristic function of the interval $[0,1]$ by such functions in the L^1-sense, we could of course dispense with them and produce the asymptotic for $\sum_{p_1+p_2+p_3=N} 1$.

10.3. A general problem

In order not to do twice the same work, let us look at the somewhat more general problem of estimating

$$(10.7) \qquad R = \sum_{n_1+n_2=h+k} \tilde{\Lambda}(n_1) \tilde{\Lambda}(n_2) u_h v_k F_N(k)$$

where $\tilde{\Lambda}(n) = \Lambda(n) F_N(n)$ and

$$(10.8) \qquad v_k = -\sum_{\substack{\ell|k \\ \ell \le L}} \mu(\ell) \operatorname{Log} \ell$$

for some $L \le N$ and general (u_h). We define

$$(10.9) \qquad S(\alpha) = \sum_n \tilde{\Lambda}(n) e(\alpha n), \quad U(\alpha) = \sum_h u_h F_N(h) e(\alpha h),$$

and get the following lemma, reminiscent of the treatment designed in (Ramaré, 1995):

Lemma 10.1. *For $D \geq 1/2$ and not more than $(\mathrm{Log}\, L)^B$ for some B, we have*

$$R = \sum_{q \leq D} \frac{\mu(q)}{\phi(q)} \sum_{a \bmod^* q} \int_{-\infty}^{\infty} S\left(\frac{a}{q} - \frac{y}{N}\right)^2 U\left(\frac{a}{q} + \frac{y}{N}\right) \hat{F}(y) dy$$

$$+ \mathcal{O}_B\left(N(ND^{-1} + L) \max_{\alpha} |U(\alpha)| (\mathrm{Log}\, N)^3\right).$$

This lemma is also reminiscent of the circle method, but the reader should notice that, $\hat{F}(y)$ having a sharp peak in 0 and decreasing rapidly as $|y|$ increases, the perturbation y/N in the exponential is a lot easier to treat than the one arising in the context of the circle method.

Proof. With the aid of (10.8), we reach

$$R = -\sum_{\ell \leq L} \mu(\ell) \mathrm{Log}\, \ell \sum_{\substack{n_1, n_2, h \\ \ell | n_1 + n_2 - h}} \tilde{\Lambda}(n_1) \tilde{\Lambda}(n_2) u_h F_N(n_1 + n_2 - h).$$

We separate variables in $F_N(n_1 + n_2 - h)$ by using the Fourier transform, getting

$$(10.10) \qquad\qquad R = \int_{-\infty}^{\infty} \hat{F}(y) R_y dy$$

where
(10.11)

$$R_y = -\sum_{\ell \leq L} \mu(\ell) \mathrm{Log}\, \ell \sum_{\substack{n_1, n_2, h \\ \ell | n_1 + n_2 - h}} \tilde{\Lambda}(n_1) \tilde{\Lambda}(n_2) u_h e\left(\frac{y(h - n_1 - n_2)}{N}\right).$$

We now detect condition $\ell | n_1 + n_2 - h$ through additive characters:

$$(10.12) \qquad \mathbb{1}_{\{\ell | n\}} = \frac{1}{\ell} \sum_{a \bmod \ell} e(na/\ell) = \frac{1}{\ell} \sum_{q | \ell} \sum_{a \bmod^* q} e(na/q).$$

Set

$$(10.13) \qquad\qquad w(q, L) = -\sum_{\substack{\ell \leq L \\ q | \ell}} \frac{\mu(\ell) \mathrm{Log}\, \ell}{\ell}.$$

We get

$$(10.14) \qquad R_y = \sum_{q \leq L} w(q, L) \sum_{a \bmod^* q} S\left(\frac{a}{q} - \frac{y}{N}\right)^2 U\left(\frac{a}{q} + \frac{y}{N}\right)$$

We propose to restrict this summation to $q \leq D$. To do so, we first notice that $|w(q, L)| \ll (\text{Log } L)^2/q$ and then proceed as follows.

$$\sum_{D<q\leq L} |w(q,L)| \sum_{a \bmod^* q} \left| S\left(\frac{a}{q} - \frac{y}{N}\right) \right|^2 \left| U\left(\frac{a}{q} + \frac{y}{N}\right) \right|$$

$$\ll (\text{Log } L)^2 \max_\alpha |U(\alpha)| \sum_{D<q\leq L} \frac{1}{q} \sum_{a \bmod^* q} \left| S\left(\frac{a}{q} - \frac{y}{N}\right) \right|^2$$

and we bound the last sum by partial summation and the large sieve inequality applied to sets of points of the form

$$\mathcal{X} = \left\{ \frac{a}{q} - \frac{y}{N} \,/\, q \leq Q, a \bmod^* q \right\}.$$

Once this reduction is done, we simplify the remaining $w(q, L)$'s, for which the prime number theorem yields

$$(10.15) \qquad w(q, L) = \frac{\mu(q)}{\phi(q)} + \mathcal{O}(2^{\omega(q)} D^{-4})$$

and such an estimate is enough. $\diamond\diamond\diamond$

10.4. Asymptotic for $\mathfrak{R}$

Let us set

$$(10.16) \qquad \mathfrak{S}_{2,2} = \prod_{p\geq 2} \left(1 + \frac{1}{(p-1)^3} \right).$$

We state formally what we establish here:

Theorem 10.3. *For any $A \geq 1$ and as N goes to infinity*

$$\mathfrak{R} = C_0 \mathfrak{S}_{2,2} N^3 + \mathcal{O}_A(N^3 (\text{Log } N)^{-A}).$$

From now on, we select $A \geq 1$.
First we note that

$$\Lambda(n) = -\sum_{d|n} \mu(d) \text{Log } d = -\sum_{\substack{d|n \\ d\leq\sqrt{N}}} \mu(d) \text{Log } d - \sum_{\substack{d|n \\ d>\sqrt{N}}} \mu(d) \text{Log } d$$

$$(10.17) \quad = \Lambda^\sharp(n) + \Lambda^\flat(n)$$

say. Since $\mu(d)$ is supposed to vary considerably in signs, we expect the last sum to contribute only to the error term. Here we follow notations of Iwaniec. We decompose $\Lambda(n_4) = \Lambda^\sharp(n_4) + \Lambda^\flat(n_4)$ in

(10.18) $$\mathfrak{R} = \sum_{n_1+n_2-n_3=n_4} \tilde{\Lambda}(n_1)\tilde{\Lambda}(n_2)\tilde{\Lambda}(n_3)\tilde{\Lambda}(n_4)$$

to split $\mathfrak{R}$ into $\mathfrak{R} = \mathfrak{R}^\sharp + \mathfrak{R}^\flat$.

Discarding $\mathfrak{R}^\flat$. We write

(10.19) $$\Lambda(n_3) = -\sum_{d|n_3} \mu(d)\,\mathrm{Log}\,d$$

so that Lemma 10.1 applies with $h = n_4$, $k = n_3$, $v_k = \Lambda(n_3)$ and $u_h = \tilde{\Lambda}^\flat(n_4)$. We choose $D = 1/2$. To handle the contribution from U, we use the following lemma from (Davenport, 1937a; Davenport, 1937b)

Lemma 10.2 (Davenport). *Uniformly in α and for every positive B, we have*

$$\left| \sum_{h \le H} \mu(h)e(h\alpha) \right| \ll_B H/(\mathrm{Log}\,H)^B.$$

This proof contains the innovation due to Vinogradov concerning the estimation of exponential sums with prime argument through a combination of sieve method and bilinear forms techniques. We do not prove this lemma here, as it is way out of our ground. But we note it also requires the use of the prime number theorem in arithmetic progressions, which we recall below.

Using this lemma, we get

Lemma 10.3. $|U(\alpha)| \ll_B N/(\mathrm{Log}\,N)^B$.

Proof. We write $k = \ell n$ and

$$U(\alpha) = -\int_{-\infty}^{\infty} \hat{F}(y) \sum_{n \le \sqrt{N}} \sum_{\sqrt{N} < n\ell \le N} \mu(\ell)(\mathrm{Log}\,\ell)e((\alpha - y/N)\ell n)dy$$

while, by partial summation, we have

$$\left| \sum_{n\ell \le N} \mu(\ell)(\mathrm{Log}\,\ell)e(\beta\ell) \right| \ll \frac{N}{n}/\mathrm{Log}^B(N/n) \ll \frac{N}{n}/\mathrm{Log}^B N.$$

The lemma follows readily.   ◇◇◇

This finally yields with $B = A + 3$

(10.20) $$\mathfrak{R}^\flat(y) = \mathcal{O}_A(N^3(\mathrm{Log}\,N)^{-A}).$$

Treating $\mathfrak{R}^\sharp$. First, we take the opportunity of this section to state a result that is so often used in this monograph.

Lemma 10.4 (The prime number theorem for arithmetic progressions). *For every constants B and C and as N goes to infinity, we have*

$$\sum_{\substack{n \leq N \\ n \equiv a[q]}} \Lambda(n) = \frac{N}{\phi(q)}(1 + \mathcal{O}_{B,C}(1/\operatorname{Log}^B N))$$

for every $q \leq \operatorname{Log}^C N$ and any a coprime to q.

We now resume the course of the proof and use Lemma 10.1 with $h = n_3$, $k = n_4$, $v_k = \Lambda^\sharp(n_4)$ and $u_h = \tilde{\Lambda}(n_3)$. We get

$$\mathfrak{R}_y^\sharp = \sum_{q \leq D} \frac{\mu(q)}{\phi(q)} \sum_{a \bmod^* q} S\left(\frac{a}{q} - \frac{y}{N}\right)^2 \overline{S^\sharp\left(\frac{a}{q} - \frac{y}{N}\right)} + \mathcal{O}(N^3(\operatorname{Log} N)^{-A})$$

where D is $(\operatorname{Log} N)^{A+3}$. At this level we can complete $S^\sharp$ by $S^\flat$ to recover S up to an affordable error term, where the reader has already understood that $S^\sharp$ (resp. $S^\flat$) stands for the trigonometric polynomial associated to $\Lambda^\sharp$ (resp. $\Lambda^\flat$).

Lemma 10.5. *We have for all $q \leq D$*

$$S\left(\frac{a}{q} + \frac{y}{N}\right) = \frac{\mu(q)N}{\phi(q)}\hat{F}(y) + \mathcal{O}(ND^{-4}(1 + |y|)).$$

Proof. First set $\check{F}_N(y) = \sum_n F_N(n)e(ny/N)$ and write

$$S\left(\frac{a}{q} + \frac{y}{N}\right) - \frac{\mu(q)}{\phi(q)}\check{F}_N(y) = \sum_n \left(\Lambda(n)e(na/q) - \frac{\mu(q)}{\phi(q)}\right) F_N(n)e(ny/N).$$

The key to this classical evaluation is to use summation by parts with respect to n. This may be surprising at start because we are trying to derive a result in $(a/q) + (y/N)$ from one in a/q. But remember this deviation has been introduced precisely to handle the size condition. This also means that we use the prime number theorem not only at size N but also for nearby values. We thus note that

$$F_N(n)e(ny/N) = -\int_n^N \Delta(t)dt$$

with $\Delta(t) = (F'(t) + 2i\pi y F(t))e(yt)/N$ which enables us to write

$$S\left(\frac{a}{q} + \frac{y}{N}\right) - \frac{\mu(q)}{\phi(q)}\check{F}_N(y) = -\int_1^N \sum_{n \leq t}\left(\Lambda(n)e(na/q) - \frac{\mu(q)}{\phi(q)}\right)\Delta(t)dt.$$

Here, we simply split the inner summation into the congruence classes of n modulo q. The n's that are not coprime with q contribute to

$$\sum_{p|q} \operatorname{Log} p \sum_{\substack{r \geq 1 \\ p^r \leq t}} 1 \leq \omega(q) \operatorname{Log} t \ll \operatorname{Log}^2 N$$

while Lemma 10.4 yields

$$\sum_{b \bmod^* q} \sum_{\substack{n \leq t \\ n \equiv \bar{b}[q]}} \Lambda(n) e(na/q) = \frac{t}{\phi(q)} \sum_{b \bmod^* q} e(ba/q) + \mathcal{O}(qN/\operatorname{Log}^B N)$$

$$= \frac{t\mu(q)}{\phi(q)} + \mathcal{O}(ND^{-4})$$

on selecting B such that $(\operatorname{Log} N)^B \geq D^5$. Gathering our estimates, we reach

(10.21) $$\qquad S\left(\frac{a}{q} + \frac{y}{N}\right) - \frac{\mu(q)}{\phi(q)} \check{F}_N(y) = \mathcal{O}(ND^{-4}).$$

Next we evaluate $\check{F}_N(y)$ in terms of $\hat{F}$ by comparing the former to an integral:

$$F(n/N)e(ny/N) = N \int_{\frac{n-1}{N}}^{n/N} F(x)e(xy)dx + \mathcal{O}(1/N).$$

The lemma follows readily.   ◇ ◇ ◇

Using the approximation given by Lemma 10.5, we infer that

(10.22) $$\qquad \mathfrak{R}_y^\sharp = \sum_{q \leq D} \frac{\mu(q)^2 N^3}{\phi(q)^3} \hat{F}(-y)^2 \hat{F}(y) + \mathcal{O}(N^3(1 + |y|)/D).$$

We shall use this bound for $|y| \leq Y$. The almost trivial bound $\mathfrak{R}_y^\sharp(D) = \mathcal{O}(N^3 \operatorname{Log} N)$ (by the large sieve inequality) suffices otherwise. This amounts to

$$\mathfrak{R}^\sharp = N^3 \int_{-\infty}^{\infty} \hat{F}(y)^2 \hat{F}(-y)^2 dy \sum_{q \leq D} \frac{\mu(q)^2}{\phi(q)^3} + \mathcal{O}\left(\frac{N^3 \operatorname{Log} Y}{D} + \frac{N^3 \operatorname{Log} N}{Y}\right)$$

in which the choice $Y = D$ is acceptable. We then simply complete the series in q. This ends the proof.

10.5. The local model

Very similar to what we did in 8.4.1, we set

(10.23) $$\qquad \varphi_q^*(n) = \frac{\mu(q)c_q(n)}{\phi(q)}(F * F)(n/N)$$

where $F * F$ denotes the usual convolution. We should expand a bit on this choice; first, one should note that it is composed of two different parts, one taking care of the arithmetic modulo q while the other one takes into account the size conditions. Second, the proper definition of the first factor should be $\mu^2(q)c_q(n)/\phi(q)^2$ as the reader will discover by computing the sum over the divisors q of d of this function, a definition that differs from our choice only by a multiplicative factor. This is irrelevant as far as the main term for given q is concerned but becomes important at the level of (10.26) where we have to add all the terms coming from $[\varphi_q^*|\varphi_{q'}^*]$ with $q' \neq q$. There, it is best to have $r(q')$ of constant mean value which explains why we divide by $\phi(q)$ in (10.23) and not by $\phi(q)^2$.

As in section 8.4.2, we get

$$[\varphi_q^*|\varphi_{q'}^*] = \frac{\mu(q)}{\phi(q)}\frac{\mu(q')}{\phi(q')} \sum_n c_q(n)c_{q'}(n)(F * F)(n/N)^2$$

and we express both Ramanujan sums in terms of divisors of q, q' and n getting

(10.24)

$$[\varphi_q^*|\varphi_{q'}^*] = \frac{\mu(q)}{\phi(q)}\frac{\mu(q')}{\phi(q')} \sum_{\substack{d|q \\ d'|q'}} d\mu(q/d)d'\mu(q'/d') \sum_{\substack{n \\ [d,d']|n}} (F * F)(n/N)^2.$$

For the innermost sum, we have

$$\sum_{\substack{n \\ [d,d']|n}} (F * F)(n/N)^2 = \int_{-\infty}^{\infty} \hat{F}(y)^2 \sum_{\substack{n \\ [d,d']|n}} (F * F)(n/N)e(-ny/N)dy$$

and using a comparison to an integral for the inner sum, we find this integral to be $C_0 N/[d, d'] + \mathcal{O}(1)$, so that

(10.25) $$[\varphi_q^*|\varphi_{q'}^*] = \frac{\mu^2(q)NC_0}{\phi(q)}\mathbb{1}_{q=q'} + \mathcal{O}(r(q)r(q'))$$

with $r(q) = \sigma(q)/\phi(q)$. This yields for fixed q:

(10.26) $$\sum_{q'} |[\varphi_q^*|\varphi_{q'}^*]| = \frac{\mu^2(q)NC_0}{\phi(q)} + \mathcal{O}(r(q)Q).$$

Let c be such that the $\mathcal{O}(r(q)Q)$ is not more in absolute value than $cr(q)Q$. We set

(10.27) $$M_q = \frac{\mu^2(q)NC_0}{\phi(q)} + cr(q)Q.$$

We further find that for $q \leq Q = (\text{Log } N)^A$, we have by expressing $c_q(n)$ in terms of $e(an/q)$ and $F * F(n/N)$ in terms of its Fourier transform

$$[r_2|\varphi_q^*] = \frac{\mu(q)}{\phi(q)} \sum_{n_1+n_2=n} \tilde{\Lambda}(n_1)\tilde{\Lambda}(n_2)(F * F)(n/N)c_q(n)$$

$$= \frac{\mu(q)}{\phi(q)} \sum_{a \bmod^* q} \int_{-\infty}^{\infty} \hat{F}(y)^2 S\left(\frac{a}{q} - \frac{y}{N}\right)^2 dy$$

$$(10.28) \qquad\qquad = \frac{\mu(q)NC_0}{\phi(q)^3} + \mathcal{O}(NQ^{-3}).$$

From which we infer

$$(10.29) \qquad \mathfrak{R} - \sum_{q \leq Q} M_q^{-1}[r_2|\varphi_q^*]^2 = \mathcal{O}(N^3/Q).$$

10.5.1. Proof of the three primes Theorem. Let N be the odd integer we want to represent. Set f_1 the characteristic function of those primes that are in the interval $]Q, N]$ (this notation represents the interval of real numbers between Q and N but where Q is excluded while N is included) and $f(n) = f_1(N - n)$. We define $[\![f|g]\!]$ as in (10.3) but with $\omega_{i,j} = 0$ and with $Q = (\text{Log } N)^{100}$. First note that

$$(10.30) \qquad r_3(N) = \sum_{n_1+n_2+n_3=N} f_1(n_3)\tilde{\Lambda}(n_1)\tilde{\Lambda}(n_2) = [f|r_2]$$

and use

$$(10.31) \qquad |[\![f|r_2]\!]|^2 \leq [\![f|f]\!][\![r_2|r_2]\!].$$

Equation (10.29) tells us that $[\![r_2|r_2]\!]$ is suitably small. It is easy to see that $[\![f|f]\!]$ is $\ll N$ so that $|[\![f|r_2]\!]|$ is small, namely

$$(10.32) \qquad |[\![f|r_2]\!]| \ll N^2/(\text{Log } N)^{50}.$$

This means in turn that $\langle f|r_2 \rangle$ approximates $r_3(N)$ sufficiently well. This leads to a quantitative version of the three primes theorem provided we compute $\langle f|r_2 \rangle$. But this is simple enough: $[r_2|\varphi_q^*]$ is given in (10.28) while

$$[f|\varphi_q^*] = \frac{\mu(q)}{\phi(q)} \sum_{n \in \mathbb{Z}} (F * F)(n/N) f(n) c_q(n)$$

$$= \frac{\mu(q)}{\phi(q)} \sum_{n_3 \in \mathbb{Z}} (F * F)((N - n_3)/N) f_1(n_3) c_q(N - n_3)$$

$$= \frac{\mu(q)}{\phi(q)} \sum_{a \bmod^* q} e\left(\frac{Na}{q}\right)$$

$$\times \int_{-\infty}^{\infty} \hat{F}^2(y) \sum_{n_3} \Lambda(n_3) e\left(\frac{-an_3}{q} + \frac{y}{N}\right) e(-y) dy$$

where we expressed $c_q(N - n_3)$ in terms of $e((N - n_3)a/q)$ and $(F * F)((N - n_3)/N)$ in terms of its Fourier transform. By now, the reader should be well acquainted with these techniques. We pursue the proof by appealing to Lemma 10.5 and finally get

$$(10.33) \qquad [f|\varphi_q^*] = \frac{\mu^2(q) N c_q(N) C_1}{\phi(q)^2} (1 + \mathcal{O}(Q^{-2}))$$

where the constant C_1 is

$$(10.34) \qquad C_1 = \int_{-\infty}^{\infty} |\hat{F}^2(y)| \hat{F}(y) e(-y/2) dy = 0.013688\ldots$$

This amounts to

$$\langle f|r_2 \rangle = N C_1 \sum_{q \leq Q} \frac{\mu(q) c_q(N)}{\phi(q)^3} + \mathcal{O}(N^2/Q)$$

and completing the summation in q, we end up with

$$(10.35) \quad \langle f|r_2 \rangle = N C_1 \prod_{p \geq 2} \left(1 + \frac{1}{(p-1)^3}\right) \prod_{p|N} \frac{p^2 - 3p + 2}{p^2 - 3p + 3} + \mathcal{O}(N^2/Q).$$

Note, and that is reassuring, that the first term vanishes if N is even. By (10.32), this expression is valid for $[f|r_2]$ which is nothing but $r_3(N)$, concluding the proof of Theorem 10.1

10.6. A slight digression

We sketch here a proof of Theorem 10.2. We are to compute

$$(10.36) \qquad V = \sum_{n} \left(r_2(n) - N\mathfrak{S}_2(n)(F * F)(n/N)\right)^2$$

with

(10.37) $$\mathfrak{S}_2(n) = C_2 \prod_{\substack{p|n \\ p \neq 2}} \frac{p-2}{p-1} = C_2 \sum_{\substack{d|n \\ (d,2)=1}} \frac{\mu^2(d)}{\phi(d)}$$

and

(10.38) $$C_2 = 2 \prod_{p \geq 3} \frac{p(p-2)}{(p-1)^2}.$$

To compute V, we expand the inner square. The first term is $\mathfrak{R}$ while the third one is trivial to estimate. As for the cross term, we write

$$\sum_n \mathfrak{S}_2(n) r_2(n) (F * F)(n/N) = C_2 \sum_{\substack{d \geq 1 \\ (d,2)=1}} \frac{\mu^2(d)}{\phi(d)} \sum_{\substack{n \geq 0 \\ d|n}} r_2(n)(F * F)(n/N).$$

In the latter expression, we notice that only the congruence classes of n_1 and n_2 modulo d intervene, with notations from (10.1). For large d, the Brun-Titchmarsh theorem is enough to show that the corresponding contribution is negligible, while for small d's, the prime number theorem in arithmetic progressions applies. The reader will finally reach

(10.39) $$V \ll_A N^3/(\text{Log } N)^A$$

meaning that most of the N summands are not more than $N^2/(\text{Log } N)^A$, and this in turns implies that for those n's, we have

(10.40) $$r_2(n) = N\mathfrak{S}_2(n)(F * F)(n/N) + \mathcal{O}_A(N/(\text{Log } N)^{A/2})$$

which is what was to be proved.

11 The Selberg sieve

In this chapter, we first present the Selberg sieve in a fashion similar to what we did up to now. In passing, we shall extend the Selberg sieve to the case of non-squarefree sifting sets, as was already done in (Selberg, 1976), but our setting will also carry through to sieving sequences and not only sets. Furthermore, this setting will enable us to compare the three different approaches: via the large sieve inequality, via local models or via the Selberg sieve.

11.1. Position of the problem

To properly set the sieve problem, one needs two objects:
   (1) A finite host sequence $\mathcal{A}$; for instance, as was the case upto now in these lectures, $\mathcal{A} = [M + 1, M + N]$.
   (2) A compact set $\mathcal{K}$, i.e. a finite collection of well-behaved – see section 2.1 – subsets $\mathcal{K}_d$ of $\mathbb{Z}/d\mathbb{Z}$.

The question is then to understand

$$(11.1) \qquad S = \{n \in \mathcal{A} \ / \ \forall d \leq D, \quad n \in \mathcal{K}_d\}$$

and, in particular, to evaluate its cardinality. We met this question already at several different places, with $\mathcal{K}_d = (\mathbb{Z}/d\mathbb{Z})^*$ the set of invertible elements modulo d to reach the prime numbers, and with $\mathcal{K}_d$ being the sets of squares modulo d to reach the (integer) squares.

11.2. Bordering system associated to a compact set

We define here another sequence of sets $(\mathcal{L}_d)_{d \geq 1}$ complementary to $(\mathcal{K}_d)$: we set $\mathcal{L}_1 = \{1\}$ and $\mathcal{L}_{p^\nu} = \mathcal{K}_{p^{\nu-1}} - \mathcal{K}_{p^\nu}$, i.e. the set of elements of $x \in \mathbb{Z}/p^\nu\mathbb{Z}$ such that $\sigma_{p^\nu \to p^{\nu-1}}(x) \in \mathcal{K}_{p^{\nu-1}}$ but that do *not* belong to $\mathcal{K}_{p^\nu}$. We further define $\mathcal{L}_d$ by "multiplicativity". It is important to note, and that is different from what happens to $\mathcal{K}$, that *we do not have* $\mathcal{L}_\ell = \mathcal{L}_d/\ell\mathbb{Z}$ if $\ell | d$. Using $\mathbb{1}_\mathcal{A}$ to denote the characteristic function of $\mathcal{A}$, our definitions imply that

$$(11.2) \quad \begin{cases} \mathbb{1}_{\mathcal{L}_d} = \prod_{p^\nu \| d} \left(\mathbb{1}_{\mathcal{K}_{p^{\nu-1}}} - \mathbb{1}_{\mathcal{K}_{p^\nu}} \right) = (-1)^{\omega(d)} \sum_{\delta | d} \mu(d/\delta) \mathbb{1}_{\mathcal{K}_\delta} \\ \mathbb{1}_{\mathcal{K}_d} = \prod_{p^\nu \| d} \left(\mathbb{1} - \mathbb{1}_{\mathcal{L}_p} - \mathbb{1}_{\mathcal{L}_{p^2}} - \cdots - \mathbb{1}_{\mathcal{L}_{p^\nu}} \right) = \sum_{\delta | d} (-1)^{\omega(d)} \mathbb{1}_{\mathcal{L}_\delta}. \end{cases}$$

A remark on why one should introduce $\mathcal{L}$: to start with, let us note that classical sieve expositions stress more on the classes that one *excludes* modulo p, than on the classes that are retained, which in our setting means that the sets $\mathcal{L}_p$ are defined first, and the sets $\mathcal{K}_p$ are usually not specified. This is so because we usually exclude few classes, i.e. $\mathcal{L}_p$ is small while $\mathcal{K}_p$ is large. This notion of *small* and *large* is in fact what led to the nomenclature "large sieve": in the example treated (see section 6.5), (Linnik, 1941) had to exclude many classes.

Introducing $\mathcal{K}_p$ allows us to get a geometrical setting, i.e. leads to a natural definition of $\mathcal{K}_d$ – while that of $\mathcal{L}_d$ is much less natural – and, in general, to smoother formulae for the main terms. However, when it comes to computing error terms, the fact that $\mathcal{L}_d$ has small cardinality in usual problems turns out to be extremely effective.

At the end of next section, we explain in terms of information this change of view point.

11.3. An extremal problem

In our presentation of the Selberg sieve, we consider the following extremal problems

$$(11.3) \quad \begin{cases} \sum_d \lambda_d^\sharp = 1 \quad , \quad \lambda_d^\sharp = 0 \quad \text{if } d \geq D \\ \text{Main term of} \displaystyle\sum_{M<n\leq M+N} \left(\sum_{d/n\in\mathcal{K}_d} \lambda_d^\sharp \right)^2 \text{minimal} \end{cases}$$

and

$$(11.4) \quad \begin{cases} \lambda_1 = 1 \quad , \quad \lambda_d = 0 \quad \text{if } d \geq D \\ \text{Main term of} \displaystyle\sum_{M<n\leq M+N} \left(\sum_{d/n\in\mathcal{L}_d} \lambda_d \right)^2 \text{minimal.} \end{cases}$$

We switch from one problem to the other using (11.2) :

$$(11.5) \quad \begin{cases} (-1)^{\omega(d)} \lambda_d = \displaystyle\sum_{d|\ell} \lambda_\ell^\sharp \quad , \quad \lambda_\ell^\sharp = \displaystyle\sum_{\ell|d} \mu(d/\ell)(-1)^{\omega(d)} \lambda_d, \\ \displaystyle\sum_{d/n\in\mathcal{L}_d} \lambda_d = \displaystyle\sum_{d/n\in\mathcal{K}_d} \lambda_d^\sharp. \end{cases}$$

Solving the first problem is very easy because $\mathcal{K}$ is multiplicatively split, and is performed via the diagonalization process of Selberg. Indeed, we write

$$\sum_{M<n\leq M+N}\left(\sum_{d/n\in\mathcal{K}_d}\lambda_d^\sharp\right)^2 = \sum_{d_1,d_2\leq D}\lambda_{d_1}^\sharp\lambda_{d_2}^\sharp\sum_{\substack{M<n\leq M+N\\n\in\mathcal{K}_{[d_1,d_2]}}}1$$

$$=\sum_{d_1,d_2\leq D}\lambda_{d_1}^\sharp\lambda_{d_2}^\sharp\frac{|\mathcal{K}_{[d_1,d_2]}|}{[d_1,d_2]}N + \text{ error term}$$

Set $\rho(d)=|\mathcal{K}_d|/d$ and let h be the solution of $1/\rho = 1 \star h$ as in (2.5). We then have

$$\sum_{d_1,d_2\leq D}\lambda_{d_1}^\sharp\lambda_{d_2}^\sharp\frac{|\mathcal{K}_{[d_1,d_2]}|}{[d_1,d_2]} = \sum_{d_1,d_2\leq D}\lambda_{d_1}^\sharp\rho(d_1)\lambda_{d_2}^\sharp\rho(d_2)(1\star h)((d_1,d_2))$$

$$=\sum_{q\leq D}h(q)\left(\sum_{q|d\leq D}\lambda_d^\sharp\rho(d)\right)^2.$$

We comment on the above relations: first we note that any two randomly chosen integers have a small gcd, so that we indeed reduce the difficulty by exchanging lcm with gcd; the next problem is still the fact that d_1 and d_2 are linked and the introduction of h is a key idea to separate them fully. Pursuing the proof, we define

(11.6)
$$y_q = \sum_{q|d\leq D}\lambda_d^\sharp\rho(d)$$

and recover the $\lambda_d^\sharp$'s from the y_q's by[1]

(11.7)
$$\rho(d)\lambda_d^\sharp = \sum_{d|q\leq D}\mu(q/d)y_q$$

which enables us to establish that

(11.8)
$$1 = \sum_d\lambda_d^\sharp = \sum_q h(q)y_q.$$

We minimize the quadratic form $\sum h(q)y_q^2$ subject to the condition (11.8). On using Lagrange multipliers, we see optimal[2] y_q's should all be equal to $1/\sum_d h(d)$ i.e. $1/G_1(D)$.

---

[1]Equation (11.6) may be seen as a linear system expressing the y_q's in terms of the $(\lambda_d^\sharp\rho(d))$'s. This system being in triangular form, the $(\lambda_d^\sharp\rho(d))$'s are uniquely determined in terms of the y_q's. The reader will check that the RHS of (11.7) verifies this system, and hence, is equal to $\lambda_d^\sharp\rho(d)$.

[2]When $h(q)$ vanishes, the corresponding value of y_q has no influence whatsoever; the corresponding λ_q will always appear with coefficient $h(q)$, The solution y_q we choose is the one that yields uniform formulae.

Gathering our results we infer (see also (18.2))
(11.9)
$$\lambda_d^\sharp = \frac{d}{|\mathcal{K}_d|} \sum_{q \leq Q/d} \mu(q)/G_1(D) \quad \text{and} \quad \lambda_d = (-1)^{\omega(d)} G_d(D)/G_1(D).$$

From the information theory point of view, going from $(\lambda_d^\sharp)$ to (λ_d) may be explained by the following remark : when writing $n \in \mathcal{K}_{p^\nu}$, we forget we already know that $n \in \mathcal{K}_{p^{\nu-1}}$; Removing this redundancy leads to $(\mathcal{L}_d)$ and to (λ_d). The reader will perhaps appreciate Lemma 2.2 better now. The L.H.S. is $G_1(D)\lambda_d^\sharp$ while the R.H.S. is its expression in terms of the λ_d's. Indeed this was how this lemma was invented.

Note that Lemma 2.3 tells us simply that $|\lambda_d| \leq 1$.

As for the cardinality of $\mathcal{S}$ (defined in (11.1)), we directly get

$$|\mathcal{S}| \leq \sum_{n \leq N} \left(\sum_{d/n \in \mathcal{K}_d} \lambda_d^\sharp \right)^2 = \sum_{n \leq N} \left(\sum_{d/n \in \mathcal{L}_d} \lambda_d \right)^2$$

(11.10)
$$\leq \frac{N}{G_1(D)} + \left(\sum_d |\mathcal{L}_d| |\lambda_d| \right)^2$$

Going from $(\lambda_d^\sharp)$ to (λ_d) is thus extremely important to reducing the error term, thanks to Lemma 2.3. Now (11.10) improves on Corollary 2.1 in that the Johnsen-Gallagher condition is no more required.

In (Selberg, 1976) and (Motohashi, 1983), the reader will find another exposition and in (Gallagher, 1974) closely related material.

Three last remarks are to be made:

(1) We do not require $\mathcal{K}$ to be squarefree.
(2) We do not require $\mathcal{K}$ to satisfy the Johnsen-Gallagher condition, contrarily to what happened in Corollary 2.1 or Theorem 2.1 . But we had access to a large sieve extension, while this result provides us with no such extension.
(3) All of what we do is valid when sieving an arbitrary sequences $\mathcal{A}$, like the sequence of primes. This only alters the definition of ρ as exposed in chapter 13. Again this is not the case of Theorem 2.1.

11.4. More on compact sets

Let $\mathcal{K}$ be a multiplicatively split compact set. We set

(11.11)
$$\psi_d(n) = \frac{d}{|\mathcal{K}_d|} \mathbb{1}_{\mathcal{K}_d}(n)$$

where the coefficient $d/|\mathcal{K}_d|$ will yield smoother formulae[3]. We have

(11.12)
$$\psi_d(n) = \sum_{q|d} \psi_q^*(n)$$

with

(11.13)
$$\psi_q^*(n) = \sum_{\substack{\delta|q \\ n\in\mathcal{K}_\delta}} \mu(q/\delta)\delta/|\mathcal{K}_\delta|.$$

It will be better to replace the condition $n \in \mathcal{K}_d$ with $n \in \mathcal{L}_d$, which we do via (11.2) and get

(11.14)
$$\psi_q^*(n) = \sum_{\substack{\ell|q \\ n\in\mathcal{L}_\ell}} (-1)^{\omega(\ell)} H(\ell,q)$$

with

(11.15)
$$H(\ell,q) = \sum_{\ell|\delta|q} \mu(q/\delta)\delta/|\mathcal{K}_\delta|.$$

Note that $H(1,q)$ is simply the function $h(q)$ we defined in (2.5) and that we did in fact already meet this function $H(\ell,q)$: Lemma 2.1 may also be written in the form

(11.16)
$$G_d(Q) = \sum_{\substack{q\leq Q \\ d|q}} H(d,q).$$

11.5. Pseudo-characters

(Selberg, 1972) introduced the notion of pseudo-characters, a notion that has proved to be most efficient in the context of log-free zero density estimates by (Motohashi, 1978). We show here that they differ from our ψ_q^*'s only by a multiplicative factor.

To do so, we follow closely chapter 1 of (Motohashi, 1983) and we start by translating his notations into our context:

- Function θ used therein and defined in (1.1.17) is in fact $\theta(q) = |\mathcal{K}_q|/q$.
- Function g defined by (1.1.18) and (1.1.21) is given in our notations by $g(q) = H(1,q) = h(q)$.
- Function Δ_q defined by the equation following (1.2.3) is $\mathbb{1}_{\mathcal{K}_q}$.

---

[3]For the reader who went through chapter 4: if $\mathcal{K}$ satisfies the Johnsen-Gallagher condition, we have $J_{\bar{d}}^{\bar{q}}(\psi_q) = \psi_d$, where J is associated with the host compact set $(\mathbb{Z}/d\mathbb{Z})_d$. The decomposition given in (11.12) is simply the one coming from (4.12). See section 9.4 for a more detailled argument. But even without knowing that $J_{\bar{d}}^{\bar{q}}(\psi_q) = \psi_d$, such an identity holds since it is proved by purely combinatorial means.

From these remarks, one easily recognizes on using (11.13) from here
and (1.2.3) from Motohashi's work that our two functions, the one that
Motohashi calls a pseudo-character and our ψ_q^*, are in fact multiples of
each other. But since this coefficient depends only on q, both notions
have the same efficiency.

The reader may consult (Graham & Vaaler, 1981) for related material.
The short paper (Elliott, 1992) shows clearly, on the example of prime
numbers, how to use these pseudo-characters to produce a sieving effect.

11.6. Selberg's bound through local models

Our aim here is to show that one can derive a bound of the same strength
as (11.10) through yet another method relying on what we termed "local
models". This last method will show clear connections between the study
of additive problems as in section 10 and this sieve method. It is a
generalization, though with a weaker remainder term, of what we did
with the Brun-Titchmarsh inequality in section 8.1.

We restrict our attention to sieving intervals, for simplicity. So our
host sequence is $[M + 1, M + N]$ and the scalar product on functions
over this interval is given by

$$(11.17) \qquad [g|h] = \sum_{M<n\leq M+N} g(n)\overline{h(n)}.$$

Let us look at the bilinear form associated to the sequence $(\psi_q^*)_{q\leq Q}$
(defined by (11.13)):

$$(11.18) \qquad \left\|\sum_q \xi_q \psi_q^*\right\|^2 = \sum_{q,q'} \xi_q \overline{\xi_{q'}} [\psi_q^*|\psi_{q'}^*].$$

On using (11.14), we infer that

$$(11.19) \qquad [\psi_q^*|\psi_{q'}^*] = \sum_{\substack{\ell|q \\ \ell'|q'}} (-1)^{\omega(\ell)} H(\ell,q)(-1)^{\omega(\ell')} H(\ell',q') \sum_{\substack{n\in\mathcal{L}_\ell \\ n\in\mathcal{L}_{\ell'}}} 1.$$

The last sum is to be computed, but the reader should note that the
condition *does not in general* reduce to $n \in \mathcal{L}_{[\ell,\ell']}$ as it would if $\mathcal{L}$ were
replaced by $\mathcal{K}$. To express this sum, we introduce a notation from (Selberg, 1976):

$$(11.20) \qquad \varepsilon(\ell,\ell') = \begin{cases} 1 & \text{if } [p|(\ell,\ell') \implies v_p(\ell) = v_p(\ell')] \\ 0 & \text{else.} \end{cases}$$

A moment reflection will reveal that

$$[\psi_q^* | \psi_{q'}^*] = \sum_{\substack{\ell|q \\ \ell'|q'}} (-1)^{\omega(\ell)+\omega(\ell')} H(\ell,q) H(\ell',q') \varepsilon(\ell,\ell') \sum_{n \in \mathcal{L}_{[\ell,\ell']}} 1$$

which becomes

$$\sum_{\substack{\ell|q \\ \ell'|q'}} (-1)^{\omega(\ell)+\omega(\ell')} H(\ell,q) H(\ell',q') \varepsilon(\ell,\ell') \left(\frac{N|\mathcal{L}_{[\ell,\ell']}|}{[\ell,\ell']} + R([\ell,\ell']) \right)$$

where $R([\ell,\ell'])$ is $\mathcal{O}^*(|\mathcal{L}_{[\ell,\ell']}|)$. The main term (i.e. the term containing N in factor) vanishes if $q \neq q'$ and equals $Nh(q)$ otherwise. So we get

$$(11.21) \qquad \left\| \sum_q \xi_q \psi_q^* \right\|^2 = N \sum_q |\xi_q|^2 h(q) + \sum_{\ell,\ell'} \varepsilon(\ell,\ell') R([\ell,\ell']) z_\ell \overline{z_{\ell'}}$$

with

$$(11.22) \qquad z_\ell = (-1)^{\omega(\ell)} \sum_{\ell|q} \xi_q H(\ell,q).$$

This study being over, we can turn to sieving questions and apply Lemma 1.2. Here f is the characteristic function of the set $\mathcal{S}$ we wish to count (defined in (11.1)). Denote its cardinality by Z. First check by using (11.13) that

$$(11.23) \qquad [f | \psi_q^*] = h(q) Z$$

so that Lemma 1.2 gives us

$$(11.24) \qquad \sum_{q \leq Q} \frac{h(q)^2 Z^2}{h(q) N} \leq Z + \sum_{\ell,\ell'} \varepsilon(\ell,\ell') R([\ell,\ell']) z_\ell \overline{z_{\ell'}}$$

with $\xi_q = Z/N$. This value of ξ_q gives

$$(11.25) \qquad z_\ell = (-1)^{\omega(\ell)} \frac{Z}{N} \sum_{q/\ell|q} H(\ell,q) = (-1)^{\omega(\ell)} Z G_\ell(Q)/N.$$

We then use $|R([\ell,\ell'])| \leq |\mathcal{L}_\ell||\mathcal{L}_\ell'|$ and $(-1)^{\omega(\ell)} G_\ell(Q)/G_1(Q) = \lambda_\ell$ to get

$$(11.26) \qquad Z \leq \frac{N}{G_1(Q)} + \frac{Z G_1(Q)}{N} \left(\sum_\ell |\mathcal{L}_\ell||\lambda_\ell| \right)^2.$$

This is to be compared with (11.10): it is slightly weaker since the coefficient $ZG_1(Q)/N$ may well be ≥ 1, though not by much. Modifying the value of ξ_d to take care of the remainder term as in Theorem 7.1 would improve on this part.

11.7. Sieve weights in terms of local models

If we look carefully at the way Lemma 1.2 is proved, we see that we approximate the characteristic function f of the set we are interested in with

$$(11.27) \qquad \sum_q \xi_q \psi_q^* = \frac{Z}{N} \sum_q \psi_q^*.$$

On the other hand, the Selberg process as we exposed it introduces the weights

$$(11.28) \qquad \sum_{d/n \in \mathcal{K}_d} \lambda_d^\sharp = \sum_d \lambda_d^\sharp \mathbb{1}_{\mathcal{K}_d}(n).$$

We now express $\mathbb{1}_{\mathcal{K}_d}$ via (11.11) and (11.12), getting

$$\sum_{d/n \in \mathcal{K}_d} \lambda_d^\sharp = \sum_d \lambda_d^\sharp \frac{|\mathcal{K}_d|}{d} \sum_{q|d} \psi_q^*(n) = \sum_q \psi_q^*(n) \sum_{q|d} \frac{|\mathcal{K}_d|}{d} \lambda_d^\sharp.$$

We readily check that

$$(11.29) \qquad \sum_{d/q|d} \frac{|\mathcal{K}_d|}{d} \lambda_d^\sharp = 1/G_1(Q)$$

hence we almost recover (11.27):

$$(11.30) \qquad \sum_{d/n \in \mathcal{K}_d} \lambda_d^\sharp = \sum_q \psi_q^*/G_1(Q).$$

The miracle here is that, even though $(Z/N) \sum_q \psi_q^*$ has been invented to approximate f, it turns out that it also majorizes this function point-wise, provided we change the first coefficient from (Z/N) to $1/G_1(Q)$. Note that (11.30) in the case of primes appears already in (Selberg, 1942), and (Selberg, 1943) and is in fact at the origin of what is now known as the Selberg sieve! It appears under the definition

$$(11.31) \qquad \Lambda_Q(n) = \sum_{q \le Q} \frac{\mu(q)}{\phi(q)} c_q(n)$$

where only the correcting factor (N/Z or $1/G_1(Q)$) is missed. Such a function has also been exploited in (Selberg, 1942), (Motohashi, 1978), (Heath-Brown, 1985), (Goldston, 1992), (Goldston, 1995), (Friedlander & Goldston, 1995), and in (Vaughan, 2003) among other places, but it is generally associated to what is sometimes known as a *Ramanujan expansion* as in (Hildebrand, 1984) and not to the notion of local models as we have introduced them here. In the above mentioned works, the function Λ is approximated by Λ_Q and contribution of the difference

$\Lambda - \Lambda_Q$ is shown to be negligible in a proper average way. One can work directly with (10.17) and replace this Λ_Q by $\Lambda^\sharp$, provided we modify slightly the bound over d there from $d \leq \sqrt{N}$ to the more general $d \leq Q$. Indeed, we have

$$(11.32) \qquad \Lambda^\sharp(n) = -\sum_{\substack{d|n \\ d \leq Q}} \mu(d) \operatorname{Log} d = \sum_{q \leq Q} w(q, Q) c_q(n)$$

where $w(q, Q)$ is defined in (10.13) and evaluated in (10.15). We obtain such an expression on using (10.12). This function $\Lambda^\sharp$ may well be a better approximation than Λ_Q in some circumstances.

11.8. From the local models to the dual large sieve inequality

Now that we have found a link between the λ_d's given by Selberg sieve and the ψ_q^*'s obtained from the point of view of *local models*, we shall get the bound given by Selberg's bound through the large sieve inequality provided the Johnsen-Gallagher condition (2.4) is satisfied. In passing, this will extend the argument of (Kobayashi, 1973) to the case of non-squarefree compact sets. Roughly speaking we proceed by expressing the function ψ_q^* in terms of additive characters modulo q. Recalling (11.13), we see that

$$\psi_q^*(n) = \sum_{\delta|q} \mu(q/\delta) \frac{\delta}{|\mathcal{K}_\delta|} \left(\sum_{b \in \mathcal{K}_\delta} \frac{1}{\delta} \sum_{c \bmod \delta} e(nc/\delta) e(-cb/\delta) \right)$$

which we modify as follows:

$$\psi_q^*(n) = \sum_{\delta|q} \mu(q/\delta) \frac{1}{|\mathcal{K}_\delta|} \sum_{b \in \mathcal{K}_\delta} \sum_{\ell|\delta} \sum_{c \bmod^* \ell} e(nc/\ell) e(-cb/\ell)$$

$$= \sum_{\ell|q} \sum_{c \bmod^* \ell} \left(\sum_{\ell|\delta|q} \mu(q/\delta) \frac{1}{|\mathcal{K}_\delta|} \sum_{b \in \mathcal{K}_\delta} e(-cb/\ell) \right) e(nc/\ell).$$

In the innermost sum, only the value of b modulo ℓ is required. On the Johnsen-Gallagher condition (2.4), a value modulo ℓ yields $|\mathcal{K}_\delta|/|\mathcal{K}_\ell|$ values of b in $\mathcal{K}_\delta$. Next the summation over δ of $\mu(q/\delta)$ is 0 if $\ell \neq q$, so that only the value $\ell = q$ remains. We have reached

$$(11.33) \qquad \psi_q^*(n) = \sum_{c \bmod^* q} \left(\frac{1}{|\mathcal{K}_q|} \sum_{b \in \mathcal{K}_q} e(-cb/q) \right) e(nc/q).$$

And of course, the coefficient that appears here is simply the Fourier coefficient of ψ_q^* ... Let us call this coefficient $\hat{\psi}_q(c/q)$. Note, however, that we have required condition (2.4) to recover this expression. We then have

$$(11.34) \quad \sum_n \left(\sum_{d/n \in \mathcal{K}_d} \lambda_d^\# \right)^2 = \sum_n \left| \sum_{q \le Q} \sum_{c \bmod^* q} \hat\psi_q(c/q) e(nc/q)/G_1(Q) \right|^2$$

where the reader will recognize the dual expression to the one studied in the large sieve inequality, i.e. (1.19). The bound there thus applies, yielding

$$(11.35) \quad \sum_n \left(\sum_{d/n \in \mathcal{K}_d} \lambda_d^\# \right)^2 \le (N + Q^2) \sum_{q \le Q} \sum_{c \bmod^* q} \left| \hat\psi_q(c/q)/G_1(Q) \right|^2.$$

Now,

$$\sum_{c \bmod^* q} \left| \hat\psi_q(c/q) \right|^2 = \frac{1}{|\mathcal{K}_q|^2} \sum_{b,b' \in \mathcal{K}_q} c_q(b - b') = \frac{1}{|\mathcal{K}_q|^2} \sum_{d|q} d\mu(q/d) \sum_{\substack{b,b' \in \mathcal{K}_q \\ b \equiv b' [d]}} 1$$

$$= \sum_{d|q} d\mu(q/d) \frac{1}{|\mathcal{K}_d|} = h(q)$$

by using once again the Johnsen-Gallagher condition. Since $\sum_q h(q) = G_1(Q)$, we have proved that

$$(11.36) \quad \sum_n \left(\sum_{d/n \in \mathcal{K}_d} \lambda_d^\# \right)^2 \le (N + Q^2)/G_1(Q)$$

namely the Selberg sieve bound for an interval through the large sieve inequality, provided (2.4) holds. This time, the error term is already evaluated and we do not have to worry whether λ_d is bounded by 1 or not.

We have thus recovered Gallagher's bound via the large sieve inequality. (Motohashi, 1983) gives a more extensive treatment of this kind of material but avoids the Johnsen-Gallagher condition. He does not get any large sieve extension, while our method gives one, but he extends the result in another direction. So, the problem of finding a large sieve extension to the sieve bound in the case of a compact set that does *not* satisfies (2.4) remains open. The reader may object that such compact sets are not common in practice; it would however enable us to gather our results in one single inequality.

Finally (Huxley, 1972b) draws on the same circle of ideas. In particular, in the case of a squarefree sieve, this author gets essentially (11.33), but starting from the weights $\sum_{d/n \in \mathcal{L}_d} \lambda_d$ instead of starting from ψ_q^* as we have done here.

12 Fourier expansion of sieve weights

The previous chapter contains an expansion of $\sum_d \lambda_d \mathbb{1}_{\mathcal{L}_d}(n)$ as a linear combination of additive characters, simply by combining (11.30) and (11.33). The theme of the present chapter is to expand similarly the sieve weights

$$(12.1) \qquad \beta_{\mathcal{K}}(n) = \left(\sum_d \lambda_d \mathbb{1}_{\mathcal{L}_d}(n) \right)^2.$$

This is indeed what is done in the case of primes in (Ramaré, 1995) and what is rapidly presented in a general context in (Ramaré & Ruzsa, 2001), equation (4.1.21). Such a material is used in (Green & Tao, 2006).

We assume throughout this chapter that $\mathcal{K}$ *is multiplicatively split and verifies the Johnsen-Gallagher condition.*

12.1. Dimension of the sieve

Up to now we have avoided to provide a general scheme to evaluate the G-functions appearing in the Selberg sieve and only drove such an evaluation in special cases. This is to encompass usual situations when the sieve is said to have a *dimension* $\kappa \geq 0$, i.e. when we have

$$(12.2) \qquad \sum_{p \leq X} (1 - |\mathcal{K}_p|/p) \operatorname{Log} p = \kappa \operatorname{Log} X + \mathcal{O}(1)$$

as well as the general case, for instance the one appearing when sieving squares as in the proof of Theorem 5.4. If (12.2) is verified and the sieve is squarefree, then Theorem 21.1 yields

$$(12.3) \qquad G_1(z) = C(\mathcal{K}) \operatorname{Log}^{\kappa} z + \mathcal{O}(\operatorname{Log}^{\kappa-1} z)$$

where $C(\mathcal{K})$ is a positive constant. We refer to (Halberstam & Richert, 1974), (Gallagher, 1974), (Iwaniec, 1980) as well as (Rawsthorne, 1982) for more details concerning sieve dimensions. We quote furthermore the paper (Vaughan, 1973) where the reader will find more difficult evaluations of $G_1(z)$.

We will simply follow the convention to say that the sieve has dimension κ whenever (12.3) holds.

12.2. The Fourier coefficients

We now define for a prime to q what will be our Fourier coefficients, namely

$$w(a/q) = \lim_{Y \to \infty} \frac{1}{Y} \sum_{n \leq Y} \left(\sum_{n \in \mathcal{K}_d} \lambda_d^{\sharp} \right)^2 e(na/q)$$

$$(12.4) \qquad = \sum_{q | [d_1, d_2]} \frac{\lambda_{d_1}^{\sharp} \lambda_{d_2}^{\sharp}}{[d_1, d_2]} \sum_{b \in \mathcal{K}_{[d_1, d_2]}} e(ab/q).$$

We shall also require the following (rather ugly) function:

$$(12.5) \qquad \rho_z(q, \delta) = \sum_{\substack{q_1 q_2 q_3 = q/(\delta, q) \\ (q_1, q_2) = (q_1, q_3) = (q_2, q_3) = 1 \\ \max(q_1 q_3 \delta, q_2 q_3 \delta) \leq z}} (-1)^{\omega(q_3)}.$$

Note that $\rho_z(q, \delta) = 1$ when $q\delta \leq z$ and vanishes when $\sqrt{q\delta} > z$ (since $\max(q_1 q_3, q_2 q_3) \geq \sqrt{q/\delta}$). Moreover we check that $|\rho_z(q, \delta)| \leq 3^{\omega(q/(\delta, q))}$. The reader should also notice that, though this function is intricate enough in its definition, it is *universal*: it does not depend on the set $\mathcal{K}$.

A third definition is required:

$$(12.6) \qquad w_q^{\sharp} = \sum_{\delta \leq z} h(\delta) \rho_z(q, \delta) / G_1(z)^2$$

where h is defined in (2.5) (see also (11.15)).

Lemma 12.1. *We have*

$$w(a/q) = \sum_{b \in \mathcal{K}_q} e(ab/q) \, w_q^{\sharp} / |\mathcal{K}_q|.$$

Proof. From (12.4), we infer

$$w(a/q) = \sum_{q | [d_1, d_2]} \frac{\lambda_{d_1}^{\sharp} \lambda_{d_2}^{\sharp}}{[d_1, d_2]} |\mathcal{K}_{[d_1, d_2]}| \frac{\sum_{b \in \mathcal{K}_q} e(ab/q)}{|\mathcal{K}_q|}$$

$$(12.7) \qquad = w_q^{\sharp} \frac{\sum_{b \in \mathcal{K}_q} e(ab/q)}{|\mathcal{K}_q|}$$

say. Replacing $\lambda^\sharp$ by its value, we get

$$G_1(z)^2 w_q^\sharp = \sum_{q|[d_1,d_2]} \frac{(d_1,d_2)}{|\mathcal{K}_{(d_1,d_2)}|} \sum_{d_1|\ell_1 \leq z} \sum_{d_2|\ell_2 \leq z} \mu(\ell_1/d_1)\mu(\ell_2/d_2)$$

$$= \sum_{\substack{\ell_1,\ell_2 \leq z}} \sum_{\substack{d_1|\ell_1, d_2|\ell_2 \\ q|[d_1,d_2]}} \frac{(d_1,d_2)}{|\mathcal{K}_{(d_1,d_2)}|} \mu(\ell_1/d_1)\mu(\ell_2/d_2)$$

i.e.

$$G_1(z)^2 w_q^\sharp = \sum_{\delta \leq z} h(\delta) \sum_{\ell_1,\ell_2 \leq z} \sum_{\substack{\delta|d_1|\ell_1 \\ \delta|d_2|\ell_2 \\ q|[d_1,d_2]}} \mu(\ell_1/d_1)\mu(\ell_2/d_2) = \sum_{\delta \leq z} h(\delta)\rho_z(q,\delta)$$

with

$$\rho_z(q,\delta) = \sum_{\ell'_1,\ell'_2 \leq z/\delta} \sum_{\substack{d_1|\ell'_1, d_2|\ell'_2 \\ q/(\delta,q)|[d_1,d_2]}} \mu(\ell'_1/d_1)\mu(\ell'_2/d_2)$$

and we now evaluate the inner sum by multiplicativity to recover our definition above. Its value is 0 as soon as there is a prime p which divides ℓ'_1 or ℓ'_2 but not $q/(\delta,q)$. Let then p be a prime such that $p^a\|\ell'_1$, $p^b\|\ell'_2$ and $p^c\|q/(\delta,q)$ with $c \geq 1$. We check successively that the value of the inner sum is 0 if $c \leq \max(a,b) - 1$, or if $c = \max(a,b) > \min(a,b) \geq 1$. Its value is 1 if $c = \max(a,b) > \min(a,b) = 0$ and -1 if $c = a = b$. We can thus write $\ell'_1 = q_1 q_3$, $\ell'_2 = q_2 q_3$ with $q/(\delta,q) = q_1 q_2 q_3$ and $(q_1,q_2) = (q_1,q_3) = (q_2,q_3) = 1$ and the value of the inner sum is $(-1)^{\omega(q_3)}$. This justifies the definition of ρ_z in (12.5).   ◇◇◇

If we have a sieve of dimension κ, then recalling (2.7) we reach

$$G_1^2(z) w_q^\sharp = G_1(z) + \mathcal{O}\big(3^{\omega(q)}(G_1(z) - G_1(z/q))\big)$$

which we combine with (12.3) to infer

(12.8) $G_1(z) w_q^\sharp = 1 + \mathcal{O}\big(3^{\omega(q)}(\operatorname{Log} q)/\operatorname{Log} z\big), \quad (q \leq z).$

Uniformy, we have the bound

(12.9) $|G_1(z) w_q^\sharp| \ll 3^{\omega(q)},$

this being a direct consequence of (12.5)–(12.6).

To conclude this part, we consider $\sum_{b \in \mathcal{K}_q} e(ab/q)$. First as an easy application of the chinese remainder theorem, we readily discover the bound

(12.10) $$\left| \sum_{b \in \mathcal{K}_q} e(ab/q) \right| \leq \prod_{p^\nu \| q} (p^\nu - |\mathcal{K}_{p^\nu}|).$$

Next if $c/M = a/q$ with $(a, q) = 1$, then note that

$$(12.11) \qquad \frac{1}{|\mathcal{K}_M|} \sum_{b \in \mathcal{K}_M} e(cb/M) = \frac{1}{|\mathcal{K}_q|} \sum_{b \in \mathcal{K}_q} e(ab/q).$$

12.3. Distribution of $\beta_{\mathcal{K}}$ in arithmetic progressions

We assume $\mathcal{K}$ is of dimension κ. We further assume that

$$(12.12) \qquad p^\nu - |\mathcal{K}_{p^\nu}| \le c p^{\nu \xi}$$

for some $c > 0$ and $\xi \in [0, \frac{1}{2}[$ which implies (see (12.7), (12.8) and (12.10))

$$(12.13) \qquad |G_1(z) w(a/q)| \ll q^{-1/2}.$$

We then get by using additive characters

$$\sum_{n \le X} \left(\sum_{n \in \mathcal{K}_d} \lambda_d^\sharp \right)^2 e(na/q) = X w_q^\sharp \frac{\sum_{b \in \mathcal{K}_q} e(ab/q)}{|\mathcal{K}_q|} + \mathcal{O}(z^2)$$

$$= \frac{X}{G_1(z)} \frac{\sum_{b \in \mathcal{K}_q} e(ab/q)}{|\mathcal{K}_q|} + \mathcal{O}\left(z^2 + \frac{X\sqrt{q}}{|\mathcal{K}_q| G_1(z) \operatorname{Log} z} \right)$$

the last equality coming from (12.8), (12.10) and (12.12). As an easy consequence and recalling (12.1), we get

$$(12.14) \qquad \sum_{\substack{n \le X \\ n \equiv b[q]}} \beta_K(n) = \frac{X \mathbb{1}_{b \in \mathcal{K}_q}}{G_1(z) |\mathcal{K}_q|} + \mathcal{O}\left(z^2 + \frac{X\sqrt{q}}{|\mathcal{K}_q| G_1(z) \operatorname{Log} z} \right).$$

12.4. Fourier expansion of $\beta_{\mathcal{K}}$

In order to have a confortable setting to evaluate $\sum_n \beta_{\mathcal{K}}(n) F(n)$, where $\beta_{\mathcal{K}}$ is defined in (12.1), we seek another expression of $\beta_{\mathcal{K}}$ as in (Ramaré, 1995). Note that

$$\beta_{\mathcal{K}}(n) = \left(\sum_{d/n \in \mathcal{K}_d} \lambda_d^\sharp \right)^2 = \sum_{d_1, d_2} \lambda_{d_1}^\sharp \lambda_{d_2}^\sharp \mathbb{1}_{\mathcal{K}_{[d_1, d_2]}}(n).$$

We now express the inner characteristic function by using additive characters and get

$$\beta_{\mathcal{K}}(n) = \sum_{d_1,d_2} \frac{\lambda_{d_1}^{\sharp} \lambda_{d_2}^{\sharp}}{[d_1,d_2]} \sum_{a \bmod [d_1,d_2]} e(an/[d_1,d_2]) \sum_{b \in \mathcal{K}_{[d_1,d_2]}} e(-ab/[d_1,d_2])$$

$$= \sum_{d_1,d_2} \frac{\lambda_{d_1}^{\sharp} \lambda_{d_2}^{\sharp}}{[d_1,d_2]} \sum_{q|[d_1,d_2]} \sum_{a \bmod {}^{*} q} e(an/q) \sum_{b \in \mathcal{K}_{[d_1,d_2]}} e(-ab/q).$$

Recalling (12.4), we see that we have reached the fundamental identity

$$(12.15) \qquad \beta_{\mathcal{K}}(n) = \sum_{q \leq z^2} \sum_{a \bmod {}^{*} q} w(a/q) e(an/q).$$

13 The Selberg sieve for sequences

The setting we developed for the Selberg sieve enables us to sieve sequences even if the compact set $\mathcal{K}$ is not squarefree, though it will still have to be multiplicatively split. The adaptation is easy enough but we record the necessary formulae and detail some examples.

13.1. A general expression

Let $(u_n)_{n\in\mathbb{Z}}$ be a weighted sequence, the weights u_n being non-negative and such that $\sum_n u_n < +\infty$. Let $\mathcal{K}$ be a multiplicatively split compact set. We assume there exists a multiplicative function $\sigma^{\sharp}$, a parameter X and a function $R_d^{\sharp}$ such that

$$\text{(13.1)} \qquad \sum_{n\in\mathcal{K}_d} u_n = \sigma^{\sharp}(d)X + R_d^{\sharp}.$$

We assume further that $\sigma^{\sharp}$ is non-negative and decreases on powers of primes (a likely hypothesis if one conceives of $\sigma^{\sharp}(d)$ as being a density), which translates into $\sigma^{\sharp}(q) \geq \sigma^{\sharp}(d)$ whenever $q|d$. Equivalently, we assume the existence of σ and R_d such that

$$\text{(13.2)} \qquad \sum_{n\in\mathcal{L}_d} u_n = \sigma(d)X + R_d$$

but the non-increasing property on chains of multiples is way less obvious to state. Switching from (13.1) to (13.2) is readily done through (11.2). There comes

$$\text{(13.3)} \qquad \begin{cases} (-1)^{\omega(d)}\sigma(d) = \displaystyle\sum_{\delta|d} \mu(d/\delta)\sigma^{\sharp}(\delta), \\[2mm] \sigma^{\sharp}(d) = \displaystyle\sum_{\delta|d}(-1)^{\omega(\delta)}\sigma(\delta). \end{cases}$$

All the analysis of section 11.3 applies, except we are to change the definition of our G-functions. First, h is the solution of

$$\text{(13.4)} \qquad \frac{1}{\sigma^{\sharp}(d)} = \sum_{q|d} h(q)$$

(compare with (2.6)), that is to say

$$\text{(13.5)} \qquad h(d) = \prod_{p^{\nu}\|\delta}\left(\frac{1}{\sigma^{\sharp}(p^{\nu})} - \frac{1}{\sigma^{\sharp}(p^{\nu-1})}\right) \geq 0.$$

Proceeding as in section 11.3, but with $\rho = \sigma^\sharp$, we get

$$(13.6) \qquad \sum_{n \in \mathcal{S}} u_n \leq \frac{X}{G_1(z)} + \sum_{d_1, d_2} \lambda_{d_1} \lambda_{d_2} R_{[d_1, d_2]},$$

with $\mathcal{S}$ defined by (11.1). Notice that we still have $|\lambda_d| \leq 1$ as in the simpler case of intervals.

13.2. The case of host sequences supported by a compact set

The two main types of sequences that we want to sieve are the sequence of prime numbers, and the one of polynomial values :

$$(13.7) \qquad \mathcal{A} = \left\{ F(n) / \ n \in [M+1, M+N] \right\}.$$

In both cases, the host sequence is supported by some multiplicatively split compact set. That is $\mathcal{U}$ for the sequence of primes, which further verifies the Johnsen-Gallagher condition (2.4); and in fact $(\mathbb{Z}/d\mathbb{Z})_d$ in the case of polynomial values! The polynomial intervenes in that our compact is of the shape $\left(F^{-1}(\mathcal{K}'_d) \right)_d$: we want n such that $F(n)$ belongs to $\mathcal{K}'_d$ for all d's below some bound.

This latter compact is wilder than it seems and does not in general verify the Johnsen-Gallagher condition: with $F = X^2$, and $\mathcal{K}' = 1 + \mathcal{U}$, the class 0 modulo p lifts to only one class modulo p^2 while all others lift to $(p-1)/2$ classes when $p \neq 2$.

We shall treat an example with the sequence of prime numbers. This sequence is carried by $\mathcal{U}$, so that in our definition of $\mathcal{L}_d$, we could restrict our attention to invertible classes, or replace $\mathcal{L}_d$ by $\mathcal{L}_d \cap \mathcal{U}_d$, as is apparent from (13.2).

We shall comment on the problem of sieving the sequence of primes with a non-squarefree sieve that would sieve out many classes in section 13.4.

13.3. On a problem of Gallagher

Let us explore and generalize a problem of (Gallagher, 1974). This will give us background information for next section, an example on how to treat the remainder term in Selberg sieve as well as an unusual application of Theorem 21.1 that we prove in the appendix. Our generalization depends on two parameters: an integer $k \geq 1$ and a polynomial F with integer coefficients. These two parameters being fixed, we consider the set defined as follows

$$\mathcal{K}'(p, F) = \{F(n) \ / \ F(n) = a_0 + a_1 p + \dots,$$

(13.8)
$$\text{with } 0 \leq a_\nu \leq p - 1, \quad a_\nu \neq 0 \text{ if } \nu < k$$
$$\text{and } n \in [M + 1, M + N]\}.$$

In the aforementioned paper, Gallagher established the upper bound

(13.9) $\left|\{n \leq N \ / \ \forall p \neq n, \ n \in \mathcal{K}'(p, X)\}\right| \leq (1 + o(1)) 2^k (k!)^2 \dfrac{N}{\text{Log}^k N}$

by using Corollary 2.1. We examine now the case $F = X^2 - 2$ and $k = 2$. We seek an upper bound for

(13.10) $\left|\{n \leq N \ / \ \forall p \neq F(n), \ F(n) \in \mathcal{K}'(p, F)\}\right|.$

Our compact set $\mathcal{K}$ is defined by split multiplicativity: modulo p^ν, it is $F^{-1}(\mathcal{K}'(p, F))$ taken modulo p^ν. First, we need the cardinality of $\mathcal{K}_q$, for which we find that

$$\begin{cases} |\mathcal{K}_2| = 1, \ |\mathcal{K}_4| = 2, \\ |\mathcal{K}_p| = p - 2, \ |\mathcal{K}_{p^2}| = p^2 - 3p + 3 \text{ if } p \equiv \pm 1[8], \\ |\mathcal{K}_p| = p, \ |\mathcal{K}_{p^2}| = p^2 - p + 1 \qquad \text{if } p \equiv \pm 3[8]. \end{cases}$$

Before proving this point, let us recall that 2 is a square modulo odd p if and only if $p \equiv \pm 1[8]$.

Proof. We only handle case $p \geq 3$. If 2 is a quadratic residue modulo p, then one should avoid its two square roots. If 2 is not a quadratic residue, then no classes are to be avoided modulo p. Let us turn to what happens modulo p^2 and consider $(a + bp)^2 - 2 = a^2 - 2 + 2abp$ with $0 \leq a, b < p$ and where $a^2 - 2$ is prime to p. If a is prime to p, then $2abp$ takes any value divisible by p by choosing b properly. If $a = 0$ then by noting that $-2 = (p - 2) + p(p - 1)$, we see that any lift of a is allowed. ◇◇◇

For higher values of ν in p^ν, there are no further constraints. We infer from the above that

$$\begin{cases} |\mathcal{L}_2| = 1, \ |\mathcal{L}_4| = 0, \\ |\mathcal{L}_p| = 2, \ |\mathcal{L}_{p^2}| = p - 3 \text{ if } p \equiv \pm 1[8], \\ |\mathcal{L}_p| = 0, \ |\mathcal{L}_{p^2}| = p - 1 \text{ if } p \equiv \pm 3[8]. \end{cases}$$

The cardinality of $|\mathcal{L}_{p^\nu}|$ vanishes when $\nu \geq 3$. Let us call $\mathcal{D}$ (resp. $\mathcal{D}'$) the set of squarefree integers whose prime factors are all $\equiv \pm 1[8]$ (resp. $\equiv \pm 3[8]$). The error arising from use of (13.6) in our problem is not more than

$$\left(\sum_{d\leq Q}|\mathcal{L}_d|\right)^2 \leq \left(\sum_{\substack{d_1d_2^2d_3^2\leq Q \\ d_1,d_2\in\mathcal{D},(d_1,d_2)=1 \\ d_3\in\mathcal{D}'}} d_2d_32^{\omega(d_2)}\right)^2 \ll (Q\operatorname{Log}^3 Q)^2.$$

As for the main term, first note that

$$\begin{cases} |h(2)| = 1, \ |h(4)| = 0, \\ |h(p)| = \dfrac{2}{p-2}, \ |h(p^2)| = \dfrac{p^2-3p}{(p-2)(p^2-3p+3)} & \text{if } p \equiv \pm1[8], \\ |h(p)| = 0, \ |h(p^2)| = \dfrac{2p-2}{p^2-p+1} & \text{if } p \equiv \pm3[8]. \end{cases}$$

We evaluate the G-function by appealing to Theorem 21.1. We note successively that

$$\sum_{\substack{p_1\leq Q \\ p_1\equiv\pm1[8]}} \frac{2\operatorname{Log} p_1}{p_1-2} = (1+o(1))2\tfrac{1}{2}\operatorname{Log} Q,$$

that

$$\sum_{\substack{p_2^2\leq Q \\ p_2\equiv\pm1[8]}} \frac{(p_2-3)p_2\operatorname{Log} p_2}{(p_2-2)(p_2^2-3p_2+3)} = (1+o(1))\tfrac{1}{2}\operatorname{Log}\sqrt{Q},$$

and finally that

$$\sum_{\substack{p_2^2\leq Q \\ p_2\equiv\pm1[8]}} \frac{(p_3-1)\operatorname{Log} p_3}{p_3^2-p_3+1} = (1+o(1))\tfrac{1}{2}\operatorname{Log}\sqrt{Q}.$$

On collecting these estimates, we find that $\kappa = 3/2$. Let us define $C = \prod_{p\geq 3} C_p$ where C_p is given by

$$C_p = \begin{cases} \left(1-\dfrac{1}{p}\right)^{3/2}\dfrac{p^2}{p^2-3p+3} & \text{when } p \equiv \pm1[8], \\ \left(1-\dfrac{1}{p}\right)^{3/2}\dfrac{p^2}{p^2-p+1} & \text{when } p \equiv \pm3[8] \end{cases}$$

so that, by taking $Q = \sqrt{N}/(\operatorname{Log} N)^4$, our cardinal defined by (13.9) is no more than

$$(1+o(1))\frac{2\sqrt{\pi}\,N}{C\,(\operatorname{Log} N)^{3/2}}.$$

13.4. On a problem of Gallagher, II

We continue to explore the preceding problem. Let $k \geq 1$ be a fixed integer. We consider again

$$\mathcal{K}'(p, X) = \{n/n = a_0 + a_1 p + \dots,$$

$$\text{with } 0 \leq a_\nu \leq p - 1, \quad a_\nu \neq 0 \text{ when } \nu < k\}.$$

While evaluating the cardinality of the L.H.S. of (13.9) in the case $F = X$, one may remark that all n belonging to the set we are interested in are prime numbers: we can thus sieve the sequence of primes instead of the one of integers. But a problem arises while controlling the error term. We are required to bound

$$(13.11) \qquad \sum_{q \leq Q} \left| \sum_{\substack{p \leq N \\ p \in \mathcal{L}_q}} \operatorname{Log} p - \frac{|\mathcal{L}_q| N}{\phi(q)} \right|$$

by $N/(\operatorname{Log} N)^{k+1}$ at least, when $Q = N^{\frac{1}{2} - \varepsilon}$. This does not follow in an immediate way from the Bombieri-Vinogradov (see Lemma 13.1 below) theorem because $|\mathcal{L}_q|$ is large. Roughly speaking, the reader will check that if $q = q_1 q_2^2 q_3^3 \cdots q_k^k$ with the q_i's being squarefree, then $|\mathcal{L}_q|$ is of order $q_2^1 q_3^2 \cdots q_k^{k-1}$. Splitting the remainder term into a contribution from each residue class and applying Hölder's inequality, we reduce the problem to bounding

$$\left(\sum_{q \leq Q} |\mathcal{L}_q|^B \max_{\ell \in \mathcal{U}_q} \left| \sum_{\substack{p \leq N \\ p \equiv \ell[q]}} \operatorname{Log} p - \frac{N}{\phi(q)} \right| \right)^{1/B} \left(\sum_{q \leq Q} \max_{\ell \in \mathcal{U}_q} \left| \sum_{\substack{p \leq N \\ p \equiv \ell[q]}} \operatorname{Log} p - \frac{N}{\phi(q)} \right| \right)^{1/A}$$

with $A^{-1} + B^{-1} = 1$. In the summation containing B, we use the Brun-Titchmarsh Theorem to dispose of the part depending on the primes. By taking A to be large and B very close to one, the Bombieri-Vinogradov Theorem (see Lemma 13.1) would allow us to prove the first factor to be not more than a power of $\operatorname{Log} Q$ if only $|\mathcal{L}_q|$ were just smaller, or if $B = 1$ were allowed.

This tantalizing problem is open.

13.5. On a subset of prime twins

Our aim here is to give an upper bound for the number of primes p not more than N that are such that $p+2$ is a prime, while $p+1$ is squarefree. The compact set $\mathcal{K}$ we choose is defined by split multiplicativity: for prime p, $\mathcal{K}_p$ is $\mathcal{U}_p \cap (\mathcal{U}_p + 2)$ while $\mathcal{K}_{p^2}$ is the set of invertibles that are not congruent to -2 modulo p and not congruent to -1 modulo p^2. For higher powers of p, $\mathcal{K}_{p^\nu}$ is defined by trivially lifting $\mathcal{K}_{p^2}$, and so will be of no interest. This yields

$$\begin{cases} |\mathcal{K}_2| = 1, \ |\mathcal{K}_4| = 1, \\ |\mathcal{K}_p| = p - 2, \ |\mathcal{K}_{p^2}| = p(p-2) - 1 = p^2 - 2p - 1 \text{ if } p \geq 3. \end{cases}$$

But now the host sequence is that of primes p weighted with a $\mathrm{Log}\, p$ each so that

$$(13.12) \qquad\qquad \sigma(d) = |\mathcal{K}_d| / \phi(d)$$

Of course $\mathcal{L}_d \cap \mathcal{U}_d$ has at most one class (class -2 modulo p and class -1 modulo p^2), implying that the error term

$$(13.13) \qquad\qquad R_d = \sum_{\substack{p \leq N \\ p \in \mathcal{L}_d \cap \mathcal{U}_d}} \mathrm{Log}\, p - \frac{|\mathcal{L}_d \cap \mathcal{U}_d| N}{\phi(d)}$$

may be controlled by

Lemma 13.1 (Bombieri-Vinogradov). *For any $B \geq 0$, there exists an $A \geq 0$ such that*

$$\sum_{q \leq Q} \max_{y \leq N} \max_{a \bmod^* q} \left| \sum_{\substack{p \leq N \\ p \equiv a[q]}} \mathrm{Log}\, p - \frac{N}{\phi(q)} \right| \ll N/(\mathrm{Log}\, N)^B$$

for $Q = \sqrt{N}/(\mathrm{Log}\, N)^A$.

Note that this "lemma" contains Lemma 10.4. By taking $B = 2$, this yields

$$\sum_{d_1, d_2 \leq D} |\lambda_{d_1} \lambda_{d_2} R_{[d_1, d_2]}| \ll N/(\mathrm{Log}\, N)^2$$

provided $D^2 = \sqrt{N}/(\mathrm{Log}\, N)^A$. As for the main term, we check that

$$\begin{cases} h(2) = 0, \ h(4) = 1, \\ h(p) = \dfrac{1}{p-2}, \ h(p^2) = \dfrac{p-1}{p^3 - 4p^2 + 3p + 2} \text{ if } p \geq 3. \end{cases}$$

Theorem 21.1 applies with $\kappa = 1$. We finally get

Theorem 13.1. *The number of primes $p \leq N$ that are such that $p + 1$ is squarefree and $p + 2$ is prime does not exceed*

$$4(1 + o(1)) \prod_{p \geq 3} \frac{p^2 - 2p - 1}{(p-1)^2} \frac{N}{\mathrm{Log}^2 N}$$

as N goes to infinity.

This bound is 4 times larger than what is conjectured but the main point here is that this bound is indeed smaller than the one one gets for prime twins (see section 21.3) by a large factor, namely

$$2 \prod_{p \geq 3} \frac{p(p-2)}{p^2 - 2p - 1} = 3.426\ldots$$

14 An overview

It is time for us to take some height and look at what we have been doing from farther away. The first approach, through the large sieve inequality, relied on an arithmetical rewriting of

$$\sum_q \sum_{a \bmod^* q} |S(a/q)|^2 \qquad \left(S(\alpha) = \sum_n u_n e(n\alpha)\right).$$

This rewriting did in fact handle the sum $W(q) = \sum_{a \bmod^* q} |S(a/q)|^2$ as one single term, and we tried to maximize it in the subsequent analysis. More precisely, whenever (u_n) vanishes outside of a given compact set, we prove a useful lower bound for this quantity.

Viewing $W(q)$ as some kind of norm (the norm of a projection onto some subspace) makes it plausible that $W(q)$ is also the scalar product of S by some function, namely the orthogonal projection of S on the proper subspace[1]. This is precisely what our *local models* φ_q^* are for: to provide a good approximation to this "projection". The case of primes is most telling: in essence, Corollary 2.1 relies on

(14.1)
$$W(q) \geq \frac{\mu^2(q)}{\phi(q)} |S(0)|^2$$

if (u_n) is carried by $\mathcal{U}$ up to at least q, while with $\varphi_q^*(n) = \mu(q)c_q(n)/\phi(q)$ defined in (8.11), we get

(14.2)
$$[(u_n)|\varphi_q^*] = \frac{\mu^2(q)}{\phi(q)} |S(0)|^2.$$

This is how local models enter the game. Note that the local models we introduced for the sums of two primes also take care of the size of the elements, so, using algebraic number theory terminology, they take care of the local contribution not only from the finite places, but also from the one at infinity.

The third viewpoint is then to try to reconstruct (u_n) from these local models, and that is exactly where the Selberg sieve comes in. We consider $C \sum_q \varphi_q^*$ with some coefficient C, and we say it ought to be an approximation of the characteristic function of our set. This scheme is also the one followed to build the function Λ_Q (see (11.32)) for the primes and is further implicit in the work of (Huxley, 1972b). But an additional uncalled for event happens here: by changing slightly the coefficient C,

---

[1]The reader who went through chapter 4 would recognize $W(q)$ as being $\|U_{\tilde{q} \to q}(\Delta_q(f))\|_q^2$, the surrounding compact set being $(\mathbb{Z}/d\mathbb{Z})_d$.

we discover that we can arrange matters so that $C \sum_q \varphi_q^*$ is exactly 1 on the set $\mathcal{S}$ we want to detect (use (11.13) and Lemma 2.1 with $d = 1$). It is expected to be of small size on the complement of $\mathcal{S}$, so, following Selberg, we replace $C \sum_q \varphi_q^*$ by its square and get an upper bound for the characteristic function of $\mathcal{S}$. This is the third aspect.

15 Some weighted sequences

Upto now, we did not investigate precisely what happens at the place at infinity. We introduced some Fourier transforms in chapter 10, and we already saw some expressions frequent in this area of mathematics in section 1.2.1. We expand all these considerations in this chapter, and, inter alia, shall provide a proof of Theorem 1.1.

The approach we follow here is due to Selberg to prove the large sieve inequality; in particular he built the function $f_{-1/2}$ given below but it turned out that Beurling had already achieved such a construction in the late 1930's without publishing. This explains why this function is now refered to as the Beurling-Selberg function.

The reader should consult the paper of (Vaaler, 1985) (see also (Graham & Vaaler, 1981)) and of (Holt & Vaaler, 1996) on which we will rely heavily. Let us note finally that the generalisation of Theorem 1.1 which we provide in Theorem 15.2 appears to be novel, as well as its corollary, Theorem 15.3.

15.1. Some special entire functions

Let $\nu > -1$ be a real number. Following (1.16) of (Holt & Vaaler, 1996) we set

$$(15.1) \qquad k_\nu(z) = k_\nu(0, z) = \frac{2\Gamma(\nu + 2)}{z} (2/z)^\nu J_{\nu+1}(z)$$

$$(15.2) \qquad = \sum_{n \geq 0} \frac{(-1)^n (z/2)^{2n} (\nu + 1)}{n!(\nu + 1)\dots(\nu + n + 1)},$$

where $J_{\nu+1}(z)$ is the Bessel function of order $\nu + 1$. Let us quote the following properties of k_ν from (Holt & Vaaler, 1996):

Lemma 15.1. *The function k_ν is even ($k_\nu(-z) = k_\nu(z)$). Its growth is controlled in vertical strips by the estimate $k_\nu(z) = \mathcal{O}\left(\exp(|\Im z|)\right)$ while on the real axis we have*

$$k_\nu(x)^2 \ll_\nu \frac{1}{(1 + |x|)^{2\nu+3}}, \quad |k_\nu(x)| \leq (\nu+1)\exp(x^2/(4(\nu+1))) \quad (x \in \mathbb{R}).$$

Finally, we also have

$$\int_{-\infty}^{+\infty} k_\nu(x)^2 |x|^{2\nu+1} dx = \Gamma(\nu + 1)\Gamma(\nu + 2)2^{2\nu+2}.$$

By a "vertical strip", we mean a set $\{z \in \mathbb{C}, a \leq \Re z \leq b\}$ for some finite a and b.

Proof.   Only the second bound is non obvious but derives easily from the Taylor expansion (15.2). $\diamond\diamond\diamond$

We deduce the following Theorem from (Holt & Vaaler, 1996):

Theorem 15.1. *There exists a real entire function ℓ_ν such that*
$$\begin{cases} \ell_\nu(z) = \mathcal{O}_\varepsilon\left(\exp\{(2+\varepsilon)|\Im z|\}\right) & \text{for any } \varepsilon > 0, \\ |\operatorname{sgn}(x) - \ell_\nu(x)| \leq k_\nu(x)^2 & (x \in \mathbb{R}). \end{cases}$$

Case $\nu = -1/2$ gives rise to the so-called Beurling-Selberg function. The reader will find an explicit expression for the functions implied in (Vaaler, 1985), together with a full presentation of the interpolation side of the problem.

Proof.   We quietly read the proof of Theorem 1 of (Holt & Vaaler, 1996), with $\xi = 0$. Equations references here refer to equations of this paper. We conclude that
$$\begin{cases} s_\nu(z, 0, 1/\pi) = \ell_\nu(0, z) - k_\nu(0, z)^2, \\ t_\nu(z, 0, 1/\pi) = \ell_\nu(0, z) + k_\nu(0, z)^2, \\ u_\nu(0, 1/\pi) = \int_{-\infty}^{\infty} k_\nu(0, x)^2 dx = \Gamma(\nu+1)\Gamma(\nu+2)2^{2\nu+2}, \end{cases}$$

on reading (5.5), (5.6) together with the comments around these equations in (Holt & Vaaler, 1996). Note that $s_\nu(z, 0, 1/\pi) = S(z)$ and $t_\nu(z, 0, 1/\pi) = T(z)$ for the proper space. The functions A_ν and B_ν are defined in (1.13) and (1.14) while the functions k_ν and ℓ_ν are defined just before the proof of Theorem 1. In particular
$$k_\nu(0, z) = \frac{K_\nu(0, z)}{K_\nu(0, 0)}, \quad \pi z K_\nu(0, z) = B_\nu(z)A_\nu(0) - A_\nu(z)B_\nu(0)$$

where the latter comes from (3.5). But $B_\nu(0) = 0$ while $A_\nu(0) = 1$, and
$$K_\nu(0, z) = \frac{B_\nu(z)}{\pi z} = \frac{\Gamma(\nu+1)}{\pi z}(2/z)^\nu J_{\nu+1}(z).$$

We find that $K_\nu(0, 0) = 1/(2\pi(\nu+1))$, so that
$$k_\nu(0, z) = \frac{2\Gamma(\nu+2)}{z}(2/z)^\nu J_{\nu+1}(z)$$

as announced. $\diamond\diamond\diamond$

Note that
$$k_{-1/2}(x) = \frac{\sin x}{x}, \quad \text{and} \quad k_{1/2}(x) = \frac{\sin x - x\cos x}{x^3/3},$$

both having value 1 at $x = 0$.

15.2. Majorants for the characteristic function of an interval

Let $\epsilon > 0$ be a real number that is fixed upto the end of the next section. Eventually, we shall let ϵ go to 0.

We consider here

$$(15.3) \qquad \chi(x) = \begin{cases} 1 & \text{if } M - \epsilon < x < M + N + \epsilon, \\ 1/2 & \text{if } x = M - \epsilon \text{ or } x = M + N + \epsilon, \\ 0 & \text{if } x \notin [M - \epsilon, M + N + \epsilon] \end{cases}$$

where M and N are two non-negative real numbers. This is the function for which we seek a well behaved majorant, where what *well behaved* exactly means will be clear from the proof below.

Let us set

$$(15.4) \qquad f_\nu(z) = \ell_\nu(z) + k_\nu(z)^2$$

and next define $b_\nu(x)$ by

$$(15.5) \quad b_\nu(x) = f_\nu(2\pi\delta(x - (M - \epsilon))) + f_\nu(2\pi\delta(M + N + \epsilon - x)),$$
$$= 2\chi(x) + f_\nu(2\pi\delta(x - (M - \epsilon))) - \mathrm{sgn}(2\pi\delta(x - (M - \epsilon)))$$
$$+ f_\nu(2\pi\delta(M + N + \epsilon - x)) - \mathrm{sgn}(2\pi\delta(M + N + \epsilon - x)).$$

By Theorem 15.1, the function b_ν is an upper bound for χ (which implies, in particular, that it is non negative) and verifies for z in $\mathbb{C}$

$$b_\nu(z) = \mathcal{O}_\varepsilon\left(\exp\{\pi\delta(2 + \varepsilon)|\Im z|\}\right) \quad \text{for any } \varepsilon > 0.$$

This bound expresses the fact that b_ν is of *exponential type* $2\pi\delta$. It holds also for $x^h b_\nu(x)$ (h non negative integer), a function that is in $\mathcal{L}_1(\mathbb{R}) \cap \mathcal{L}_2(\mathbb{R})$, both results provided $2\nu + 2 > h$.

Lemma 15.2. *Let $\nu > -1$. Let R be a polynomial of degree $< 2\nu + 2$ and $\alpha \in \mathbb{R}/\mathbb{Z}$. We have, if $|\alpha| \geq \delta$,*

$$\sum_{n \in \mathbb{Z}} R(n) b_\nu(n) e(n\alpha) = 0.$$

Furthermore

$$\sum_{n \in \mathbb{Z}} R(n) b_\nu(n) = \int_{-\infty}^{+\infty} R(t) b_\nu(t) dt.$$

Proof. We write $R = \sum r_h X^h$. The Poisson summation formula gives

$$\sum_{n \in \mathbb{Z}} R(n) b_\nu(n) e(n\alpha) = \sum_{m \in \mathbb{Z}} \sum_h r_h \hat{b}^{(h)}(m - \alpha)/(2i\pi)^h.$$

Every term on the R.H.S. vanishes when $|\alpha| \geq \delta$ by the Paley-Wiener Theorem and our remark that $n^h b_\nu(n)$ is of exponential type $2\pi\delta$ when $h < 2\nu + 2$.   ◇◇◇

15.3. A generalized large sieve inequality.

Let us start with some preliminary material on polynomials. Let $Q \in \mathbb{C}[X]$ of degree $\leq 2\nu + 1$ and define $Q^*(X) = Q((X - M + \epsilon)/N)$. Lemma 15.2 yields

$$\sum_{n\in\mathbb{Z}} Q^*(n)b_\nu(n) = N \int_{-\infty}^{+\infty} Q(t)b_\nu(Nt + M + \epsilon)dt$$

$$= N\int_0^1 Q(t)dt + \mathcal{O}^*\left(\frac{1}{2\pi\delta}\int_{-\infty}^{+\infty} k_\nu^2(t)\Big|Q\Big(\frac{t}{2\pi N\delta}\Big) + Q\Big(\frac{1-t}{2\pi N\delta}\Big)\Big|dt\right)$$

$$= N\int_0^1 Q(t)dt + \mathcal{O}^*\left(\delta^{-1}\rho_\nu(Q, 2\pi\delta N)\right)$$

say, where $\rho_\nu(Q, \xi)$ is an upper bound for

$$\frac{1}{2\pi}\int_{-\infty}^{+\infty} k_\nu^2(t)\Big(\big|Q(t/\xi) + Q((1-t)/\xi)\big|\Big)dt.$$

We define further

(15.6) $$Q^\flat(x) = \sum_h |q_h|x^{-h} \text{ when } Q(x) = \sum_h q_h x^h.$$

The following lemma provides us with a manageable upper bound for $\rho_\nu(Q, \xi)$.

Lemma 15.3. *We have $\rho_\nu(Q, \xi) \leq \rho_\nu^\flat Q^\flat(\xi)$ where*

$$\rho_\nu^\flat = \max_{0 \leq h \leq 2\nu+1} \frac{1}{2\pi}\int_{-\infty}^{+\infty} k_\nu^2(t)|t^h + (1-t)^h|dt.$$

Moreover $\rho_\nu^\flat \leq \frac{3}{2}(2\nu+2)^{2\nu+2}$ for $\nu \geq -1/2$ and more precisely $\rho_{-1/2}^\flat = 1$.

Proof.   To give an upper bound for ρ_ν, note that $|t^h + (1 - t)^h| \leq (1 + 2^{2\nu+1})\max(|t|^{2\nu+1}, 1)$ if $h \leq 2\nu + 1$. Using Lemma 15.1, we get

$$\rho_\nu^\flat \leq \frac{1 + 2^{2\nu+1}}{2\pi}\Big\{2(\nu + 1)^2 \exp(1/(2(\nu + 1))) + \Gamma(\nu + 1)\Gamma(\nu + 2)2^{2\nu+2}\Big\}.$$

It is then easy to numerically verify the upper bound, since we can control what happens for large ν by using

$$\Gamma(x) \leq \sqrt{2\pi x}(x/e)^x \exp(1/(12x)), \quad (x > 0).$$

see (Abramowitz & Stegun, 1964) equation (6.1.38). As a matter of fact, we have the stronger bound $\rho_\nu^\flat \leq \frac{1}{6}(2\nu + 2)^{2\nu+2}$ for $\nu \geq 1/2$ and $\rho_{1/2}^\flat \leq 3.6$ ◇◇◇

Let us now set

$$(15.7) \quad S_Q(\alpha) = \sum_{n=M}^{M+N} a_n Q^*(n) e(n\alpha) = \sum_{n=M}^{M+N} a_n Q\left(\frac{n-M}{N}\right) e(n\alpha).$$

The following theorem generalizes Theorem 1.1, which we recover on taking $\nu = -1/2$ and $\mathcal{Q} = \{1\}$ since the previous lemma gives $\rho_{-1/2}^\flat = 1$.[1]

Theorem 15.2. *Let $\mathcal{Q}$ be a finite set of polynomials of degree $\leq \nu + 1/2$ and orthonormal for the scalar product $\int_0^1 P_1(t)\overline{P_2(t)}dt$. Let $\mathcal{X}$ be a δ-well spaced set of points of $\mathbb{R}/\mathbb{Z}$. We have*

$$\sum_{Q\in\mathcal{Q}}\sum_{x\in\mathcal{X}} |S_Q(x)|^2 \leq \|S\|_2^2\left(N + \delta^{-1}\rho_\nu^\flat \sum_{Q\in\mathcal{Q}} Q^\flat(2\pi\delta N)^2\right).$$

In this Theorem, ϵ is taken to be 0 and $\rho_\nu^\flat$ is defined in Lemma 15.3.

Proof. We have

$$\Sigma = \sum_{Q\in\mathcal{Q}}\sum_{x\in\mathcal{X}} |S_Q(x)|^2 = \sum_{n=M}^{M+N} a_n \sum_{Q\in\mathcal{Q}}\sum_{x\in\mathcal{X}} \overline{S_Q(x)}Q^*(n)e(nx)$$

to which we apply Cauchy's inequality. We get

$$\Sigma^2 \leq \|S\|_2^2 \sum_{n\in\mathbb{Z}} b_\nu(n) \sum_{Q_1,Q_2\in\mathcal{Q}}\sum_{x,y\in\mathcal{X}} \overline{S_{Q_1}(x)}S_{Q_2}(y)Q_1^*(n)\overline{Q_2^*(n)}e(n(x-y))$$

which is not more than

$$\|S\|_2^2\left\{N \sum_{Q\in\mathcal{Q}}\sum_{x\in\mathcal{X}} |S_Q(x)|^2 + \delta^{-1}\rho_\nu^\flat \sum_{x\in\mathcal{X}}\left(\sum_{Q\in\mathcal{Q}} |S_Q(x)|Q^\flat(2\pi\delta N)\right)^2\right\}.$$

We use Cauchy's inequality on the inner square and get

$$\Sigma \leq \|S\|_2^2\left\{N + \delta^{-1}\rho_\nu^\flat \sum_{Q\in\mathcal{Q}} Q^\flat(2\pi\delta N)^2\right\}$$

as required. Note that $\mathcal{Q}$ has cardinality at most $\nu + 1$. We finally let ϵ tend to 0. ◇◇◇

---

[1]The reader might wonder why we have N here instead of $N-1$ as was announced. This is because our interval is slightly different, from $M+0$ to $M+N$ instead of from $M+1$ to $M+N$.

15.4. An application

One may wonder whether this result is stronger than the classical large sieve inequality or not. Well, in fact, it is essentially equivalent, at least if $2\pi\delta N$ is bounded below away from 0, and for the following reason. First, it contains this inequality; on the other side, the modifications introduced by the Q^* enables us to localize n essentially in intervals of size $N/(\nu + 3/2)$. However, one could first split our interval in smaller pieces and we apply on these the classical large sieve inequality; we get this way bounds of the same strength as the one above. Nonetheless, this inequality has some interesting consequences, as we shall see below.

For $\mathcal{Q}$, we can take a modification of the Legendre polynomials:

$$(15.8) \quad \mathcal{Q} = \{L_m(2X - 1), m \leq \nu + 1/2\}, \quad L_m = \frac{1}{2^m m!}\frac{d^m}{dx^m}(1 - x^2)^m.$$

We have

$$(15.9) \qquad L_0 = 1, \ L_1 = x, \ L_2 = (3x^2 - 1)/2, \ L_3 = (5x^3 - 3x)/2.$$

Restricting our attention to $\nu = 1/2$, we readily get the following Theorem, which generalizes Corollary 2.1.

Theorem 15.3. *Assume $\mathcal{K}$ is multiplicatively split and verifies the Johnsen-Gallagher condition (2.4). Let f be the characteristic function of those integers of the interval $[M, M + N]$ that belongs to $\mathcal{K}_q$ for all $q \leq Q \leq 2\sqrt{N}$. We set $Z = \sum_n f(n)$. We have*

$$Z + \left|\sum_n \left(\frac{2n}{N} - 1\right)f(n)\right|^2 /Z \leq (N + 20Q^2)/G_1(Q).$$

Proof.   The proof is straightforward and only varies from the one of Corollary 2.1 in that we use Theorem 15.2 with $\nu = 1/2$ instead of Theorem 1.1 and use the set $\mathcal{Q}$ defined in (15.8). Our hypothesis $Q \leq 2\sqrt{N}$ ensures us that $2\pi\delta N \geq 2$, so that

$$\sum_{Q\in\mathcal{Q}} Q^\flat(2\pi\delta N)^2 \leq 1 + (1 + 2\tfrac{1}{2})^2 = 5.$$

The reader will easily draw the required conclusion. $\diamond\diamond\diamond$

Compared to Corollary 2.1, we lose the factor 20, a fact that is irrelevant for most applications, but gain the term $\left|\sum_n \left(\frac{2n}{N} - 1\right)f(n)\right|^2/Z$. It contributes if the un-sifted elements (the n's with $f(n) = 1$) accumulate more on $[0, N/2]$ or on $[N/2, N]$. For the Brun-Titchmarsh inequality, it means that we could beat the constant 2 *if* such a case were to happen ...

Similar type of results have recently been studied by (Coppola & Salerno, 2004).

15.5. Perfect coupling

The proof above has several interesting features, but a main one is to provide us with a weighted sequence that is perfectly behaved in arithmetic progressions: the sequence $(b_\nu(n))$ is a majorant for the sequence of integers in the interval $[M, M + N]$ and verifies

$$\sum_{n \in \mathbb{Z}} b_\nu(n) e(na/q) = 0 \quad \text{if } (a, q) = 1 \text{ and } 1 < q \leq 1/\delta$$

so that, if we set $B = \sum_{n \in \mathbb{Z}} b_\nu(n)$, we get

(15.10) $$\forall q \leq \delta^{-1}, \qquad \sum_{\substack{n \in \mathbb{Z} \\ n \equiv c[q]}} b_\nu(n) = B/q$$

where the main feature is that *no error term* occurs. By taking ν large enough, we can also be sure that $b_\nu(n)$ decreases rapidly enough. Concerning its derivative, the Paley Wiener Theorem ensures that the derivative of f_ν defined in (15.4) is indeed bounded in terms of ν, so that

(15.11) $$|b'_\nu(x)| \ll_\nu \delta.$$

Such an inequality proves that b_ν does not vary much over intervals of size not more than δ^{-1}.

We shall see in the next chapter how such a weighted sequence may be used to considerably simplify the study of the hermitian product derived from a local system.

16 Small gaps between primes

In this chapter, we show how the perfectly well distributed weighted sequence $(b_\nu(n))_n$ built in the preceding chapter can be used to simplify the analysis of the hermitian product stemming from a local system. We show furthermore that the key point of Bombieri & Davenport's proof concerning small differences between primes is in fact contained in Lemma 1.2 and 1.1.

16.1. Introduction

Small differences between primes are a choice subject between additive and multiplicative number theory. To show this difference is infinitely often equal to 2 is nothing else than the prime twin conjecture. We consider here a much more modest aim and show that $(p_{n+1}-p_n)/\operatorname{Log} p_n$ is infinitely often ≤ 0.5 and even a bit better, where (p_n) is the sequence of primes.

The prime number Theorem tells us that there are asymptotically $x/\operatorname{Log} x$ prime numbers up to x, so that the mean difference is $\operatorname{Log} x$, which implies that $(p_{n+1}-p_n)/\operatorname{Log} p_n$ is infinitely often $\leq 1+\varepsilon$ for every $\varepsilon > 0$.

We set

$$(16.1) \qquad \Lambda_1 = \liminf \frac{p_{n+1} - p_n}{\operatorname{Log} p_n}.$$

In 1940 Erdös was the first one to go beyond $\Lambda_1 \leq 1$ in (Erdös, 1940) by showing that $\Lambda_1 \leq 57/59$. He of course did not use the Bombieri-Vinogradov Theorem (which was proved only in 1965). This result has then been improved upon, in (Rankin, 1947) and then again in (Rankin, 1950) by plugging in sieve results. A further improvement was achieved in (Ricci, 1954) where the inequality $\Lambda_1 \leq 15/16$ is proved. The second major step is due to (Bombieri & Davenport, 1966) which establishes that $\Lambda_1 \leq 0.467$, this time using the Bombieri-Vinogradov Theorem.

(Huxley, 1973) started another round of improvements by introducing combinatorial arguments. In 1977, he finally got $\Lambda_1 \leq 0.443$.

This part of the story ends up with (Maier, 1988) who employed his now famous matrix method and improved all previous results by an $e^{-\gamma}$ factor. In particular $\Lambda_1 \leq 0.249$.

(Goldston et al., 2005) is a major breakthrough in this area. Methods used therein are not foreign to what is exposed here but are overall too new and shifting to be part of this book. The reader will find several preprints on the Arxiv server.

We prove here that $\Lambda_1 \leq 1/2$ by using the setting developed till here. In particular we do not require any circle method. It would be an easy task to improve on this bound, and we indicate in a last section how to achieve this. Note that if we were to avoid the Bombieri-Vinogradov Theorem, the base method developed here would yield $\Lambda_1 \leq 1$, that could also be improved into $\Lambda_1 < 1$. This time only the prime number theorem in arithmetic progressions would be used so that we could get *effective* results and even explicit ones. Furthermore, the simplicity of the approach renders it usable for a large variety of sequences.

Throughout this chapter we shall use

$$(16.2) \qquad \mathfrak{S} = 2 \prod_{p \geq 3} (1 - (p-1)^{-2}) \quad , \quad \mathfrak{S}(j) = \mathfrak{S} \prod_{2 < p | j} \frac{p-1}{p-2}.$$

16.2. Some preliminary material

Lemma 16.1. *There exists a positive constant c_0 such that every interval of length at least $c_0/\phi(q)$ contains at least a point a/q with $(a, q) = 1$.*

Proof. Let I be an interval of length $q/\phi(q)$ in $[1, q]$. Fix some $u > 0$. The number of points divisible by a prime factor $\geq q^u$ is at most $1 + q^{1-u}\phi(q)^{-1}$ and the number of such primes it as most $1/u$. Thus the number of points in I that are coprime to all the prime factors of q less than $z = q^u$ for a small enough u is, by Brun's sieve (Theorem 2.1 of Chapter 2 of (Halberstam & Richert, 1974), on taking $\kappa = b = 1$ and $\lambda = 0.1$) , at least

$$(16.3) \qquad \frac{2}{3}|I|\frac{\phi(q)}{q} - (1 + q^{1-u}\phi(q)^{-1})/u \geq \frac{1}{2}|I|\frac{\phi(q)}{q}$$

when q is large enough, say $q \geq q_0$, and u small enough. When q is small, there is at least one such point in $[1, q]$. Hence the result with $c_0 = \max\{\phi(q), q < q_0\}$. $\diamond\diamond\diamond$

Lemma 16.2. *We have*

$$\sum_{a \bmod^* q} \left| \sum_{1 \leq m \leq M} g(m)e(ma/q) \right|^2 \gg \left(1 - \mathcal{O}(M^2/\phi(q)^2)\right) \phi(q) \sum_m |g(m)|^2$$

where the constants implied in the $\mathcal{O}$- and $\gg$-symbols do not depend on g nor on q.

Proof. Put $S(u) = \sum_{1 \leq m \leq M} g(m)e(mu)$ and set $\delta = c_0/\phi(q)$ the value given by Lemma 16.1. We then have

$$|S(a/q) - S(u)| \leq 2\pi|u - a/q| \sum_{1 \leq m \leq M} |mg(m)|$$

and thus

$$|S(a/q)|^2 \geq |S(u)|^2 - 4\pi|u - a/q| \sum_{1 \leq m \leq M} |mg(m)| \sum_m |g(m)|.$$

Integrating this inequality yields

$$|S(a/q)|^2 \geq \delta^{-1} \int_{a/q - \delta/2}^{a/q + \delta/2} |S(u)|^2 \, du - 2\pi\delta M \left(\sum_{1 \leq m \leq M} |g(m)| \right)^2$$

from which the result follows easily.   ◇◇◇

16.3. The actors and their local approximations

Let $2J + 2$ be the minimum of $p' - p$ when $p' > p \geq N$: we are required to bound J from above. Let K be an integer which we assume prime in the range $\frac{1}{10} \text{Log } N \leq K \leq \text{Log } N$. The simplest application will take $K \simeq J$. We also assume N large enough, and in particular $K \neq 2$.

For each $j \in [1, K]$, we consider

$$(16.4) \quad f^{(j)}(n) = \begin{cases} \text{Log}(n + 2j) & \text{if } n + 2j \text{ is a prime in } \in [N, 2N], \\ 0 & \text{else,} \end{cases}$$

as well as $f = \sum_j f^{(j)}$. We shall approximate these functions modulo q by

$$(16.5) \quad \varphi_q^{(j)} = \mathbb{1}_{(n+2j,q)=1} \sqrt{b(n)}$$

where $b(n) = b_{-1/2}(n)$ is described in the previous chapter, at the level of equation (15.6), related to the parameter $\delta = Q^{-2}$. We shall chose the parameter Q later on: it regulates in (16.8) below the size of the sifting set on moduli. This weighted sequence is introduced so as to have (16.10). Since having a proper majorant at the end point of the interval $[N, 2N]$ is not important, we take $\epsilon = 0$ and find that

$$\sum_{n \in \mathbb{Z}} b(n) = \hat{b}(0) = N + Q^2.$$

To $\varphi_q^{(j)}$, we also associate (see (8.10))

$$(16.6) \quad \varphi_q^{(j)*}(n) = \frac{\mu(q)c_q(n + 2j)}{\phi(q)} \sqrt{b(n)}.$$

We further set

(16.7) $$\varphi_q^* = \sum_{1 \le j \le K} \varphi_q^{(j)*}.$$

We have just been talking about *approximating* but we still have to specify for which norm ... A gap we fill in the next two subsections.

The hermitian product. Our main local system is given by $(\varphi_q^*)_{q \in \mathcal{Q}}$ for

(16.8) $$\mathcal{Q} = \{q \le Q, \mu^2(q) = 1, q \ne K, 2K\}.$$

The parameter Q can be taken as $\sqrt{N}/\mathrm{Log}^A N$ for a sufficiently large A; this is the level to which we shall sieve and is forced by our use of the Bombieri-Vinogradov Theorem (see Lemma 13.1). If in this theorem, one could reach moduli till Q^θ say, with $\theta > 1/2$, then we would get $\Lambda_1 \le 1 - \theta$ by following our proof (or Bombieri & Davenport's original one). We take simply $Q = \sqrt{N} \exp(-\sqrt{\mathrm{Log}\, N})$, to avoid having to see which power A we will need choose. This gives rise to the hermitian product

(16.9) $$\langle h|g \rangle = \sum_{q \in \mathcal{Q}} [h|\varphi_q^*] \overline{[g|\varphi_q^*]} / [\varphi_q^*|\varphi_q^*]$$

as in (10.2), since

(16.10) $$[\varphi_q^*|\varphi_{q'}^*] = 0 \qquad (\forall q \ne q' \le Q).$$

Such a relation of course simplifies a great deal of our work. Since we could dispense with it in this proof, it cannot be considered as being essential. The reader should however keep in mind when dealing with such a problem of this possibility. Looking back on the way we initially proved Theorem 1.1, we see that the factor $\sqrt{F(n)}$ in (1.8) had exactly the same role.

We need an *apriori* lower bound for $[\varphi_q^*|\varphi_q^*]$.

Lemma 16.3. *When $q \in \mathcal{Q}$, then $K^3[\varphi_q^*|\varphi_q^*] \gg N/\phi(q)$.*

Proof. Let us remark that

(16.11) $$[\varphi_q^*|\varphi_q^*] = \sum_{1 \le j,k \le K} \frac{\mu(q)^2}{\phi^2(q)} c_q(2(j-k))(N + Q^2)$$

(16.12) $$= \frac{\mu(q)^2(N + Q^2)}{\phi^2(q)} \sum_{a \bmod^* q} \left| \sum_{1 \le j \le K} e(2aj/q) \right|^2.$$

When $q \geq cK$ for a large enough c, then Lemma 16.2 yields the result. When $q \leq cK$, we restrict the sum over a to $a = 1$. We have

$$\sum_{1 \leq k \leq K} e(2k/q) = e(2/q)\frac{1 - e(2K/q)}{1 - e(2/q)}$$

which is at least $1/q$ if $q \nmid 2K$ so that $K^3[\varphi_q^*|\varphi_q^*] \gg N/\phi(q)$ in this case. The case $q = 2$ is readily worked out. Note that for $q = K$ or $q = 2K$ (forbidden by our definition of $\mathcal{Q}$), the above norm indeed vanishes. $\diamond\diamond\diamond$

Replacing $f^{(j)}$ by its local approximation $(\varphi_q^{(j)})_q$. The main Theorem on which the proof really relies is the following one. In its proof, we show that the Bombieri-Vinogradov Theorem enables us to approximate $f^{(j)}$ by its local approximation $(\varphi_q^{(j)})_q$. The proof then splits into two parts: showing that the hypothesis of this theorem are met (what we call *apriori estimates*), and computing the resulting arithmetical expressions, a part that is tedious but with no real difficulties.

Theorem 16.1. *Let $(\alpha_q)_{q \in \mathcal{Q}}$ be a sequence of complex numbers, with $|\alpha_q| \leq 2^{\omega(q)}$. We have*

$$\left|\sum_{q \in \mathcal{Q}} \alpha_q[f^{(j)} - \varphi_q^{(j)}|\varphi_q^*]\right| \ll N/\mathrm{Log}^{100} N$$

uniformly in $j \leq K$.

Proof. We first check that

$$[f^{(j)}|\varphi_q^{(k)*}] = \frac{\mu(q)}{\phi(q)} \sum_{b \bmod q} c_q(b + 2k) \sum_{\substack{n \equiv b[q] \\ n+2j \in \mathcal{P}}} \mathrm{Log}(n + 2j)\sqrt{b(n)}$$

$$= \frac{\mu(q)}{\phi(q)} \sum_{a \bmod q} c_q(a + 2k - 2j) \sum_{\substack{m \equiv a[q] \\ m \in \mathcal{P}}} \mathrm{Log}\, m\, \sqrt{b(m - 2j)}$$

by setting $m = n + 2j$, and $b + 2j = a$. If we were to approximate here the sum over m by $N/\phi(q)$ when a is prime to q, then we would require such an approximation for all a modulo q: the Bombieri-Vinogradov Theorem would not be enough to conclude. We can however reduce this approximation to essentially a single progression as follows:

$$[f^{(j)}|\varphi_q^{(k)*}] = \frac{\mu(q)}{\phi(q)} \sum_{d|q} d\mu(q/d) \sum_{\substack{a \bmod q, \\ a+2(k-j)\equiv 0[d]}} \sum_{\substack{m\equiv a[q] \\ m\in\mathcal{P}}} \mathrm{Log}\, m \sqrt{b(m-2j)}$$

$$(16.13) \qquad = \frac{\mu(q)}{\phi(q)} \sum_{d|q} d\mu(q/d) \sum_{\substack{m\equiv -2(k-j)[d] \\ m\in\mathcal{P}}} \mathrm{Log}\, m \sqrt{b(m-2j)}.$$

So for every d having some prime factor in common with $2(k-j)$ (that is, for all $d \neq 1$'s when $j = k$), the contribution from the sum over m is very small, in fact $\mathcal{O}(\mathrm{Log}\, d)$, while otherwise we may approximate this sum by $N/\phi(d)$. The overall error term thus introduced is bounded via the Bombieri-Vinogradov Theorem (see Lemma 13.1; removing the $\sqrt{b(m-2j)}$ is no problem because j being small enough, the derivative of the function is properly controlled by the fact that Q is small, see (15.11)) by $\mathcal{O}(N/\mathrm{Log}^{100} N)$. As for the main term, it is

$$\frac{\mu(q)N}{\phi(q)} \sum_{\substack{d|q \\ (d,2(j-k))=1}} \frac{d\mu(q/d)}{\phi(d)}$$

which exactly equals $[\varphi_q^{(j)}|\varphi_q^{(k)*}]N/(N+Q^2)$, as expected.   ◇◇◇

16.4. Computation of some hermitian products

We first establish an apriori upper bound for $[f^{(j)}|\varphi_q^{(k)*}]$.

Lemma 16.4. *We have* $|[f^{(j)}|\varphi_q^{(k)*}]| \ll 2^{\omega(q)} N/\phi(q)$.

Proof. We use (16.13) and the Brun-Titchmarsh inequality to get

$$|[f^{(j)}|\varphi_q^{(k)*}]| \ll \frac{\mu(q)^2}{\phi(q)} \sum_{d|q} dN/\phi(d) \ll 2^{\omega(q)} N/\phi(q)$$

as demanded.   ◇◇◇

We next compute the required scalar products $\langle f^{(j)}|f\rangle$.

Lemma 16.5. *Uniformly in* $1 \leq j \leq K$, *we have* $\langle f^{(j)}|f\rangle = (2 + o(1))KN$.

Proof. We write

$$\langle f^{(j)}|f^{(k)}\rangle = \sum_{q\in Q}[f^{(j)}|\varphi_q^*]\overline{[f^{(k)}|\varphi_q^*]}/[\varphi_q^*|\varphi_q^*]$$

in which we first replace $f^{(j)}$ by $\varphi_q^{(j)}$ by using Theorem 16.1. Hypothesis are met by appealing to Lemma 16.3 and 16.4. As a second step we replace $f^{(k)}$ by $\varphi_q^{(k)}$ and reach

$$\langle f^{(j)}|f^{(k)}\rangle = \sum_{q\in Q}[\varphi_q^{(j)}|\varphi_q^*]\overline{[\varphi_q^{(k)}|\varphi_q^*]}/[\varphi_q^*|\varphi_q^*] + \mathcal{O}(N/\mathrm{Log}^{97} N).$$

We sum this expression over k and reach

$$\langle f^{(j)}|f\rangle = \sum_{q\in Q}[\varphi_q^{(j)}|\varphi_q^*] + \mathcal{O}(N/\mathrm{Log}^{96} N).$$

Now we have as in (16.11)

$$[\varphi_q^{(j)}|\varphi_q^*] = [\varphi_q^{*(j)}|\varphi_q^*] = \sum_{1\le k\le K}\frac{\mu(q)^2}{\phi^2(q)}c_q(2(j-k))(N+Q^2).$$

Summing over q, we readily recognize

$$(N+Q^2)\Big(\sum_{k\ne j}\mathfrak{S}(j-k) + \mathcal{O}(Q^{-1}+K^{-1}) + G(Q)\Big)$$

where the $\mathcal{O}(K^{-1})$ is here to take care of the condition $q\ne K, 2K$. Next write

(16.14)
$$\mathfrak{S}(j-k) = \mathfrak{S}\sum_{\substack{d|j-k\\(d,2)=1}}\frac{1}{\phi_2(d)}$$

with $\phi_2(d) = \prod_{p|d}(p-2)$, getting

$$\sum_{k\ne j}\mathfrak{S}(j-k) = \mathfrak{S}\sum_{\substack{d\le K,\\(d,2)=1}}\frac{1}{\phi_2(d)}\sum_{k\equiv j[d],k\ne j}1$$

$$= \mathfrak{S}\sum_{\substack{d\le K,\\(d,2)=1}}\frac{1}{\phi_2(d)}\Big(\frac{K}{d}+\mathcal{O}(1)\Big) = (2+o(1))K.$$

The lemma follows readily.   ◇◇◇

16.5. Final argument

We take for K the largest prime not more than J, where $2J + 2$ is the minimum of $p' - p$ when $p' > p \geq N$. Due to the smoothing $b(n)$, Lemma 1.2 simply reads

$$(16.15) \qquad\qquad \langle f|f \rangle \leq [f|f].$$

We continue by expanding $[f|f]$: the products $[f^{(j)}|f^{(k)}]$ vanish by hypothesis when $k \neq j$. On the L.H.S., since we are able to compute the relevant products by Lemma 16.5, we show that they do not vanish, and thus this will force J to be small enough. More precisely we have

$$(16.16) \qquad \sum_{1 \leq j \leq K} 2KN + \sum_{1 \leq j \leq K} \tfrac{1}{2} N \operatorname{Log} N \leq \sum_{1 \leq j \leq K} (1 + o(1)) N \operatorname{Log} N$$

from which we infer that $2K \leq (\tfrac{1}{2} + o(1)) \operatorname{Log} N$ as required.

   Let us sketch a method for improving on this bound. Take K somewhat larger than J but still $\leq 2J$. The same inequality gives

$$2K^2 N \leq (1 + o(1)) \tfrac{1}{2} KN \operatorname{Log} N + \sum_{|j-k|>J} \sum_{\substack{N<p,p'\leq 2N, \\ p-p'=2(j-k)}} \operatorname{Log} p \operatorname{Log} p'.$$

We can control the right hand side by using the sieve bound

$$\sum_{\substack{N<p,p'\leq 2N, \\ p-p'=2(j-k)}} \operatorname{Log} p \operatorname{Log} p' \leq 4(1 + o(1)) N \mathfrak{S}(2(j - k))$$

which is 4 times larger than what is expected to be true. A proof of this upper bound is provided in subsection 21.3, at least in the case $j - k = 1$, but the general case is not much more difficult. We readily check by appealing to (16.14) that

$$\begin{cases} \forall j \leq K - J, \quad \displaystyle\sum_{k \geq j+J} \mathfrak{S}(2(j - k)) = 2(K - J - j) + o(K), \\[2mm] \forall j \geq J, \quad \displaystyle\sum_{k \leq J-j} \mathfrak{S}(2(j - k)) = 2(j - J) + o(K), \end{cases}$$

so that

$$\sum_{|j-k|>J} \sum_{\substack{N<p,p'\leq 2N, \\ p-p'=2(j-k)}} \operatorname{Log} p \operatorname{Log} p'$$

$$\leq 8N \left(\sum_{j \leq K-J} (K - J - j) + \sum_{K \geq j \geq J} (j - J) \right) + o(N \operatorname{Log}^2 N)$$

which we finally bound by $8N(K - J)^2 + o(N \operatorname{Log}^2 N)$. Set $K/J = \theta$ and $(\operatorname{Log} N)/J = \lambda$. We choose θ in such a fashion that first

$$4\theta^2 > \theta\lambda + 16(\theta - 1)^2$$

and second so that λ is maximal. We take $\theta = 2/\sqrt{3}$ and reach $\Lambda_1 \leq (2 + \sqrt{3})/8 = 0.466 \cdots < 1/2$ as announced.

17 Approximating by a local model

It is high time we show in a somewhat general setting how to approximate a given weighted sequence by a local model. Let us start with such a sequence $(f(n))_n$ together with an additional function ψ_∞ (which will take care of the size constraints), for which we assume the following bound:

$$(17.1) \qquad \sum_{q \leq D} \max_{a \bmod q} \left| \sum_{n \equiv a[q]} f(n)\psi_\infty(n) - f_q(a)X/q \right| \leq E$$

for some parameters D, E, X and $(f_q)_q$. The Bombieri-Vinogradov Theorem falls within this framework with ψ_∞ being the characteristic function of *real* numbers $\leq N$ and $E = N/(\mathrm{Log}\, N)^A$, together with $D = \sqrt{N}/(\mathrm{Log}\, N)^B$ for some $B = B(A)$; then $f(n) = \Lambda(n)$ and $f_q = q\mathbb{1}_{\mathcal{U}_q}/\phi(q)$, and finally $X = N$. Note that the function f_q that appears is precisely the one we used as a local model for the primes. The parameter X is here for homogeneity and could be dispensed with, simply by incorporating it in f_q. However, in usual applications, X will be here to treat the dependence on the size, i.e. the contribution of the infinite place, while f_q will be independent of it and only accounts for the effect of the finite places. We shall need some properties of these f_q's, namely:

$$(17.2) \qquad \forall d|q, \forall a \bmod d, \quad J_{\frac{d}{q}}^{\tilde{q}} f_q = f_d.$$

This equation may look unpalatable, but here is an equivalent formulation:

$$(17.3) \qquad \forall d|q, \quad f_d(a)/d = \sum_{\substack{b \bmod q \\ b \equiv a[q]}} f_q(b)/q$$

where it is maybe easier to consider f_q/q as one function (the *density*, as in (13.1) and (13.2)).

We often need an individual upper bound for each of this remainder term. This is not fundamental, and the end of the proof can be made to work with a large amount of variants, but usual sequences do verify this additional hypothesis: we assume that there exist $A \geq 1$ and a constant C such that, for all $q \leq Q$, we have

$$(17.4) \qquad \left| \sum_{n \equiv a[q]} f(n)\psi_\infty(n) - f_q(a)X/q \right| \leq CX A^{\Omega(q)}/q.$$

Let us turn next toward the base scalar product we use, where again we seek some generality. Let $\mathcal{K}'$ be a multiplicatively split compact set, $\mathcal{L}'$ its bordering system and

$$(17.5) \qquad \beta_{\mathcal{K}'}(n) = \left(\sum_{n \in \mathcal{L}'_d} \lambda_d \right)^2$$

be the associated Selberg's weights (see (12.1)), where $\lambda_d = 0$ whenever $d > z$ a parameter at our disposal. The scalar product we consider is

$$[\beta_{\mathcal{K}'} f | g] = \sum_{n \geq 1} \beta_{\mathcal{K}'}(n) f(n) \overline{g(n)}$$

over functions belonging to $\ell^2(\mathbb{N})^1$. This way of denoting the scalar product has the advantage of making the dependance in $\mathcal{K}'$ appear explicitly.

We are also to use another multiplicatively split compact set $\mathcal{K}$ satisfying the Johnsen-Gallagher condition, together with its bordering system $\mathcal{L}$ and define ψ_q^* as in (11.13). We further select the same function ψ_∞ as in the beginning, and define

$$(17.6) \qquad \psi_{q,\infty}^* = \psi_q^* \psi_\infty.$$

Some comments on this additional ψ_∞ are called for. We can expect to be able to prove (17.1) for a whole bunch of functions ψ_∞, like all the ones of type $g(n/N)$ for some smooth g with compact support. We could have phrased our hypothesis in these terms, then taken for ψ_∞ in (17.6) a function verifying proper conditions, and in due course, we would have discovered that it is enough to have both functions equal one to another. This is indeed the process that is followed in applications, but the exposition is simpler the way we took – albeit the need for this remark!

One way to get a global grasp of the family $(f_q)_q$ is to consider a (multiplicatively) large modulus M: by which we mean a modulus divisible by all the q's that intervene. Then f_M is enough to reproduce all f_q's, simply by $f_q = J_{\tilde{q}}^{\tilde{M}} f_M$. The reader may have doubts as to the very existence of such an f_M, but remembering the Fourier decomposition we produced, we may simply take

$$f_M = \sum_{q \leq D} L_{\tilde{M}}^{\tilde{q}} U_{\tilde{q} \to q} f_q.$$

Usually, we have at our disposal a smoother expression, like in the case of primes where $f_M = M \mathbb{1}_{\mathcal{U}_M}/\phi(M)$ is a good choice. Such an expression is in no way unique since only its orthonormal projections "modulo q" for all $q \leq Q$ are of use.

---

[1]By which we design the set of sequences $(f(n))$ such that $\sum_{n \geq 1} |f(n)|^2$ is finite.

If we have at our disposal such a modulus M that is divisible by every integer $\leq z^2$, then $\beta_{\mathcal{K}'}(n)$ has a well defined meaning for $n \in \mathbb{Z}/M\mathbb{Z}$.

Let finally $(\alpha_q)_{q \leq Q}$ be a sequence of complex numbers for which we do not assume anything. However, we think of α_q as being bounded by a divisor function. We are to understand $\sum_{q \leq Q} \alpha_q[f|\psi_q^*]$, for which the following theorem is the main key.

Theorem 17.1. *Let M be an integer divisible by every integer $\leq D = z^2 Q$. All other parameters are described above. We have*

$$\left| \sum_{q \leq Q} \alpha_q \left([\beta_{\mathcal{K}'} f|\psi_{q,\infty}^*] - X[\beta_{\mathcal{K}'} f_M | L_{\bar{M}}^{\bar{q}} \psi_q^*]_M \right) \right| \leq \left(X B E \sum_{q \leq Q} |\alpha_q|^2/q \right)^{1/2}$$

with

$$B = C \sum_{d \leq D} \left(\sum_{d = [\ell, d_1, d_2]} |\mathcal{L}_\ell| |\mathcal{L}'_{d_1}| |\mathcal{L}'_{d_2}| \left(\sum_{\ell | q \leq Q} q H(\mathcal{K}, \ell, q)^2 \right)^{1/2} \right)^2 A^{\Omega(d)}/d.$$

We expect B to evaluate to some power of $\mathrm{Log}\, D$ and this is readily done given some decent hypothesis on $\mathcal{K}$ and $\mathcal{K}'$. The function $H(\mathcal{K}, \ell, q)$ is indeed the one defined by (11.15) but we have added an explicit dependence in $\mathcal{K}$ to avoid confusion.

Proof. We start from (11.14) and write

$$[f|\psi_{q,\infty}^*] = \sum_{\ell | q} (-1)^{\omega(\ell)} H(\ell, q) \sum_{n \in \mathcal{L}_\ell} \beta_{\mathcal{K}'}(n) f(n) \psi_\infty(n).$$

Next, we write

$$\beta_{\mathcal{K}'}(n) = \sum_{\substack{d_1, d_2 \\ n \in \mathcal{L}'_{d_1} \cap \mathcal{L}'_{d_2}}} \lambda_{d_1} \lambda_{d_2}$$

so that

$$[f|\psi_{q,\infty}^*] = \sum_{\ell | q} (-1)^{\omega(\ell)} H(\ell, q) \sum_{d_1, d_2 \leq z} \lambda_{d_1} \lambda_{d_2} \sum_{n \in \mathcal{L}_\ell \cap \mathcal{L}'_{d_1} \cap \mathcal{L}'_{d_2}} f(n) \psi_\infty(n).$$

The most inner sum bears on residue classes modulo $[\ell, d_1, d_2]$ and we expect the set $\mathcal{L}_\ell \cap \mathcal{L}'_{d_1} \cap \mathcal{L}'_{d_2}$ to be small enough. We introduce the remainder term

$$\max_{a \bmod q} \left| \sum_{n \equiv a[q]} f(n) \psi_\infty(n) - X f_q(a)/q \right| = r_q^*$$

and first study the main term arising from this approximation. It equals

$$X \sum_{\ell | q} (-1)^{\omega(\ell)} H(\ell, q) \sum_{d_1, d_2 \leq z} \lambda_{d_1} \lambda_{d_2} \sum_{a \in \mathcal{L}_\ell \cap \mathcal{L}'_{d_1} \cap \mathcal{L}'_{d_2}} f_{[\ell, d_1, d_2]}(a)/[\ell, d_1, d_2].$$

We go one huge step up and write it as

$$X \sum_{\ell | q} (-1)^{\omega(\ell)} H(\ell, q) \sum_{d_1, d_2 \leq z} \lambda_{d_1} \lambda_{d_2} \sum_{c \bmod M} \frac{f_M(c)}{M} \mathbb{1}_{\mathcal{L}_\ell}(c) \mathbb{1}_{\mathcal{L}'_{d_1}}(c) \mathbb{1}_{\mathcal{L}'_{d_2}}(c)$$

which we fold back into

$$X \sum_{\ell | q} (-1)^{\omega(\ell)} H(\ell, q) \sum_{c \bmod M} \beta_{\mathcal{K}'}(c) f_M(c) \mathbb{1}_{\mathcal{L}_\ell}(c) / M$$

and finally into

$$X \sum_{c \bmod M} \beta_{\mathcal{K}'}(c) f_M(c) \psi_q^*(c) / M.$$

We used a number of usual imprecisions during these steps: we should have written $L_{\tilde{M}}^{\tilde{d_1}} \mathbb{1}_{\mathcal{L}'_{d_1}}(c)$ instead of $\mathbb{1}_{\mathcal{L}'_{d_1}}(c)$ since this latter function has arguments in $\mathbb{Z}/d_1\mathbb{Z}$ and not in $\mathbb{Z}/M\mathbb{Z}$... A similar remark holds for $\mathbb{1}_{\mathcal{L}'_{d_2}}(c)$ and for $\psi_q^*(c)$.

We handle the remainder term in a most straightforward way, with the firm belief that the cardinality of $\mathcal{L}_\ell \cap \mathcal{L}'_{d_1} \cap \mathcal{L}'_{d_2}$ as a subset of $\mathbb{Z}/[\ell, d_1, d_2]\mathbb{Z}$ will be small enough: we simply majorize it by $|\mathcal{L}_\ell||\mathcal{L}'_{d_1}||\mathcal{L}'_{d_2}|$ where the first (resp. second, resp. third) one is a cardinality as a subset of $\mathbb{Z}/\ell\mathbb{Z}$ (resp. $\mathbb{Z}/d_1\mathbb{Z}$, resp. $\mathbb{Z}/d_2\mathbb{Z}$). The remainder is then at most

$$\sum_{d \leq D} \left(\sum_{\substack{d=[\ell, d_1, d_2], \\ d_1, d_2 \leq z, \\ \ell \leq Q}} |\mathcal{L}_\ell||\mathcal{L}'_{d_1}||\mathcal{L}'_{d_2}| \sum_{\ell | q \leq Q} |H(\ell, q)||\alpha_q| \right) r_d^*.$$

Since we prepare this Theorem for the case when $\mathcal{L}_\ell$ is small, we expect $H(\ell, q)$ to behave like $1/q$ up to a divisor function. This motivates the next line:

$$\sum_{\ell | q \leq Q} |H(\ell, q)||\alpha_q| \leq \sum_{\ell | q \leq Q} q H(\ell, q)^2 \sum_{q \leq Q} |\alpha_q|^2 / q,$$

which most probably looses a factor $1/\ell$ that does not matter much. We use Cauchy's inequality and hypothesis (17.4) to conclude.   ◇ ◇ ◇

18 Selecting other sets of moduli

Concerning the moduli, we used mainly the simple condition $d \leq z$, while everything we do is valid with a condition $d \in \mathcal{D}$ for some divisor closed set[1]. Usual sets are $\{d \leq z\}$, or the set of integers $\leq z$ and with prime factors belonging to some sets (like prime to 2 or bounded by some y), or with a bounded number of prime factors.

Let us record some formulae. We set

$$(18.1) \qquad G_d(\mathcal{D}) = \sum_{\substack{\delta \in \mathcal{D} \\ [d,\delta] \in \mathcal{D}}} h(\delta)$$

and following the method shown chapter 11 and 13, we get (see (11.9))

$$(18.2) \qquad \lambda_d^{\sharp} = \frac{d}{|\mathcal{K}_d|} \frac{\sum_{q/dq \in \mathcal{D}} \mu(q)}{G_1(\mathcal{D})} \quad \text{and} \quad \lambda_d = (-1)^{\omega(d)} \frac{G_d(\mathcal{D})}{G_1(\mathcal{D})}.$$

These expressions may yield some surprises. For instance, on taking for $\mathcal{D}$ the set of those primes that are $\leq z$, to which we add the element 1, we find that $\lambda_p = -h(p)/G_1(\mathcal{D})$. This can be extremely small, and does not appear as a modification of $\mu(p) = -1$ anymore! However $\lambda_p^{\sharp}$ may be used as such.

18.1. Sieving by squares

While studying squarefree numbers, we should consider only square moduli, and a large sieve inequality related to these moduli comes in handy, as in (Konyagin, 2003). In this direction, Baier and Zhao, together or independantly got several bounds as in (Zhao, 2004a), (Zhao, 2004b), (Baier, 2006) and (Baier & Zhao, 2005). Their latest result to date states that for $N, Q > 0$ and any $\varepsilon > 0$, we have

$$(18.3) \quad \sum_{q \leq Q} \sum_{a \bmod^* q^2} \left| \sum_{n \leq N} u_n e(na/q^2) \right|^2$$

$$\ll_{\varepsilon} (NQ)^{\varepsilon} \left(Q^3 + N + \min(N\sqrt{Q}, \sqrt{N}Q^2) \right) \sum_n |u_n|^2.$$

If we were to use the large sieve inequality for all moduli $q^2 \leq Q^2$, we would get the upper bound $(N + Q^4) \sum_n |u_n|^2$. We do not give any

---

[1]A *divisor closed set* is a set such that every positive divisor of an element of this set still belongs to this set. In particular 1 always belongs to such a set.

further details here but refer to (Baier & Zhao, 2005) for background informations as well as other bounds. We are to stress that what could be the best possible inequality in (18.3) is not known. In particular, lower results are missing.

(Granville & Ramaré, 1996) used such a large sieve inequality to study the distribution of squarefree binomial coefficients; we already mentionned the recent work (Konyagin, 2003), but the most beautiful application appears in (Baier & Zhao, 2006b). The authors first derive in (Baier & Zhao, 2006a) a Bombieri-Vinogradov type theorem, by following the now classical lines of (Bombieri *et al.*, 1986) – that we also followed in section 5.5 –. From there, they prove:

Theorem 18.1 (Baier & Zhao). *Let $\varepsilon > 0$. There exist infinitely many primes p that can be written in the form $p = \ell m^2 + 1$ with $\ell \ll_\varepsilon p^{5/9+\varepsilon}$.*

An old and much sought-after conjecture of (Hardy & Littlewood, 1922) asserts that $\ell = 1$ is admissible.

18.2. A warning

In section 8-9 of (Bombieri, 1987), the sum

$$\sum_{n \in \mathcal{A}} \beta_{\mathcal{K}'}(n) \left(\sum_{d|n+2} a_d \right)$$

is being studied, with no constraints on the a_d's and where $\mathcal{A}$ is a host sequence. In the application given there, the host sequence is the one of primes, while $a_1 = 1$, $a_p = -1$ when p is a prime $\leq z$ and a_d is otherwise 0. The idea is to show that this sum tends to infinity: since $\sum_{d|n+2} a_d$ can only be positive if $n + 2$ has all its prime factors $> z$, this will show that there are infinitely many primes p such that $p + 2$ has no prime factors $\leq z$. The temptation here is to replace condition $d|n + 2$ by $n \in \mathcal{L}_d$ with $\mathcal{L}$ being the bordering system associated with $\mathcal{U} - 2$, and to consider a set of moduli $\mathcal{D}$ restricted to the primes $\leq z$ and to 1. Then we would like to take for a_d the associated λ_d: however, we have just seen that these ones are too small! The good candidates are in fact the $\lambda_d^\sharp$, but we are to modify the compact set, and take for $\mathcal{K}$ ... precisely the previous bordering system! That is to say $\mathcal{K}_p = \{-2\}$. The next step would be to appeal to (11.30):

$$(18.4) \qquad \sum_{d/n \in \mathcal{K}_d} \lambda_d^\sharp = \sum_q \psi_q^* / G_1(z).$$

We could combine Theorem 17.1 together with the Bombieri-Vinogradov Theorem (see Lemma 13.1) and be done. But our change of compact set has another consequence: the associated bordering system is not small anymore, and the error term given in Theorem 17.1 is unsuitable. We leave the reader at this level!

19 Sums of two squarefree numbers

To illustrate further how we may handle additive problems with the material we have presented, we prove the following Theorem. Note that we freely use chapter 4 in the sequel.

Theorem 19.1. *Every large enough integer may be written as the sum of two squarefree integers. Furthermore the number $r(N)$ of ways of representing N is this manner verifies:*

$$r(N) = \mathfrak{S}(N)N + \mathcal{O}_\varepsilon(N^{2/3+\varepsilon})$$

for every $\varepsilon > 0$, where

$$\mathfrak{S}(N) = \frac{6}{\pi^2} \prod_{p|N} \left(1 + \frac{p + c_{p^2}(N)}{p^2(p^2 - 1)}\right).$$

The function c_{p^2} is again the Ramanujan sum, see (8.12).

This theorem is originally due to (Evelyn & Linfoot, 1931). A simplified proof was later given by (Estermann, 1931). (Brüdern & Perelli, 1999) gave a quantitatively better version, but we are interested here in the manner. Of particular interest is the fact that a path initially devised to get *an upper bound* can be used to get a *lower bound* (as in chapter 10). Several features of the method will also be exhibited. The problem is furthermore interesting in that the set of moduli we use is different from the usual one. We take

(19.1) $\mathcal{Q} = \{q_1 q_2^2, \ \mu^2(q_1 q_2) = 1, \ q_1 \le Q_1, q_2 \le Q_2\}$

for some parameters Q_1 and Q_2 that we shall choose later to be, respectively, $N^{1/6}$ and $N^{1/3}$. We also set $Q = \max(Q_1, Q_2)$.

19.1. Sketch of the proof

Let us consider the functions $f(n) = \mu^2(n)$ and $g(n) = \mu^2(N - n)$ defined on integers $\le N$. On using the canonical hermitian product on this space, our number of representations reads

(19.2) $R(N) = \sum_{N=n_1+n_2} \mu^2(n_1)\mu^2(n_2) = [f|g].$

To compute this scalar product, we shall use again the remark we made in section 10.1 and approximate it by a $\langle f|g \rangle$ which we still have to define. We first need to work out local models for f and g. We shall then use a local system $(\varphi_i)_i$ that will take care of the sequence of local models of f and g, and that implies, in particular, that our arguments are going to be symmetrical in f and g.

19.2. General computations

To detect squarefree integers[1] we use the classical formula

$$(19.3) \qquad \mu^2(n) = \sum_{d^2 | n} \mu(d)$$

and extend it to negative values of the argument n by setting $\mu^2(-n) = \mu^2(n)$. Let us set

$$(19.4) \qquad \gamma_q(c) = \begin{cases} 0 & \text{if } \exists p^2 | q, c \equiv 0 \ [p^2] \\ \dfrac{6}{\pi^2} \prod_{p|q} \dfrac{p^2}{p^2 - 1} \prod_{\substack{p\|q \\ c \equiv 0[p]}} \left(1 - \dfrac{1}{p}\right) & \text{else.} \end{cases}$$

Lemma 19.1. *For $m = 0$ or $m = N$, we have for every $\varepsilon > 0$*

$$\sum_{\substack{n \equiv c[q] \\ n \leq N}} \mu^2(n - m) = (N/q)\gamma_q(c - m) + \mathcal{O}_\varepsilon(q^\varepsilon \sqrt{N/q}).$$

The reader will notice by comparing with (4.20) that the computed quantities are nothing else than $\Delta_q(f)/q$ when $m = 0$ (since $\mu^2(-n) = \mu^2(n)$) and $\Delta_q(g)/q$ if $m = N$.

Proof. Let us write $q = q_1 q'$ with q_1 being squarefree, q' being such that $p|q' \implies p^2|q'$ and, of course $(q_1, q') = 1$. If $m - c$ is divisible by a square that also divides q, then $\sum_{n \equiv c[q]} \mu^2(n - m) = 0$. Else we use the representation (19.3) to get

$$\sum_{\substack{n \equiv c[q] \\ n \leq N}} \mu^2(n - m) = \sum_d \mu(d) \sum_{\substack{n \leq N \\ n \equiv m - c[q] \\ d^2 | n}} 1 = \sum_{\substack{d_1 | q_1 \\ (d_3, q) = 1}} \mu(d_1 d_3) \sum_{\substack{n \leq N \\ n \equiv m - c[q] \\ d_1^2 d_3^2 | n}} 1.$$

---

[1]The characteristic function of their set is μ^2.

We use an asymptotic for the inner sum and conclude readily:

$$\sum_{\substack{n\equiv c[q]\\ n\le N}} \mu^2(n-m) = N \sum_{\substack{d_1|q_1\\ (d_3,q)=1\\ 0\equiv m-c[d_1]\\ q_1 d_1 q' d_3^2 \le N}} \frac{\mu(d_1 d_3)}{q_1 d_1 q' d_3^2} + \mathcal{O}\left(\sum_{d_1|q_1} \sqrt{\frac{N}{qd_1}}\right)$$

$$= (N/q)\gamma_q(c-m) + \mathcal{O}(q^\varepsilon \sqrt{N/q}).$$

◇◇◇

We define γ_q^* by $\gamma_q^*(c) = \sum_{d|q} \mu(q/d)\gamma_d(c)$. This implies that $\gamma_q = \sum_{d|q} \gamma_d^*$, so that each γ_d^* is indeed the orthonormal projection of γ_q on $\mathfrak{M}(d)$ (see the comment following Lemma 4.2 for a definition of this set). Let us define

(19.5)
$$t(q) = \prod_{p|q} \frac{-1}{p^2 - 1}.$$

Now we have

Lemma 19.2. *If q is cubefree then $\gamma_q^*(c) = 6t(q)c_q(c)/\pi^2$, while if q has a cubic factor > 1, then $\gamma_q^*(c) = 0$.*

Proof. We write $q = q_1 q_2^2 q''$ where q_1, q_2 and q'' are pairwise coprime, q_1 and q_2 are squarefree and, if $p|q''$, then $p^3|q''$. We have

$$\gamma_q^*(c) = \frac{6}{q\pi^2} \prod_{\substack{p|q''\\ p^2\nmid(c,q)}} \left(\frac{p^2}{p^2-1} - \frac{p^2}{p^2-1}\right) \prod_{\substack{p|q''\\ p^2|(c,q)}} (0)$$

$$\times \prod_{\substack{p|q_2\\ p^2|(c,q)}} \left(\frac{-p^2}{p^2-1}\left(1-\frac{1}{p}\right)\right) \prod_{\substack{p|q_2\\ p\|(c,q)}} \frac{p^2}{p^2-1}\left(1-1+\frac{1}{p}\right) \prod_{\substack{p|q_2\\ p\nmid(c,q)}} \frac{p^2}{p^2-1}(1-1)$$

$$\times \prod_{p|q_1} \left(\frac{p^2}{p^2-1}-1\right) \prod_{p|(q_1,c)} \left((p^2-1)\left(\frac{p^2}{p^2-1}\left(1-\frac{1}{p}\right)-1\right)\right).$$

Thus $\gamma_q^*(c) = 0$ if $q'' \ne 1$, and

(19.6) $\quad \gamma_q^*(c) = \dfrac{6}{q\pi^2} \displaystyle\prod_{p|q} \frac{1}{p^2-1} \prod_{p|(q_1,c)} (1-p) \times \begin{cases} p - p^2 & \text{if } p^2|(c,q_2^2), \\ p & \text{if } p\|(c,q_2^2), \\ 0 & \text{else.} \end{cases}$

Some more work yields the claimed expression.   ◇◇◇

19.3. The hermitian product

We must first embark onto some general considerations. Lemma 19.1 shows that $c \mapsto N\gamma_q(c)$ is a good approximation for $\Delta_q(f)/q$, while $N\theta_q : c \mapsto N\gamma_q(N-c)$ is a good one for $\Delta_q(g)/q$, where Δ_q is defined in (4.20). However, by a local model, we mean a function over $[1, N]$ and not modulo q. This distinction is important to define the hermitian product, so we need to lift both functions to this set. We consider

$$
(19.7) \qquad \begin{aligned} \nabla_q : \mathscr{F}(\mathbb{Z}/q\mathbb{Z}) &\longrightarrow \mathscr{F}([1, N]) \\ h &\mapsto \nabla_q(h) : [1, N] \longrightarrow \mathbb{C} \\ &\qquad\qquad x \mapsto h(x \bmod q) \end{aligned}
$$

which verifies

$$
(19.8) \qquad [\Delta_q(h_1)|h_2]_q = [h_1|\nabla_q(h_2)],
$$

justifying again our scaling in the definition of Δ_q. Note further that

$$
(19.9) \qquad \forall d|q, \quad \nabla_q(L_{\tilde{q}}^{\tilde{d}}(h)) = \nabla_d(h),
$$

both properties stated with obvious notations.

This part being settled, we need to attend to a second problem before the proof can unfold quietly. We need an orthogonal system modulo q that takes both functions $N\gamma_q$ and $N\theta_q$ into account, or more precisely encompasses their orthogonal projections $N\gamma_q^*$ and $N\theta_q^*$. Let us first note that

$$
(19.10) \quad \|\gamma_q^*\|_q^2 = \|\theta_q^*\|_q^2 = \left(\frac{6t(q)}{\pi^2}\right)^2 \frac{1}{q} \sum_{a \bmod q} |c_q(a)|^2 = \left(\frac{6t(q)}{\pi^2}\right)^2 \phi(q).
$$

Since $[\gamma_q^*|\theta_q^*]_q = (6t(q)/\pi^2)^2 c_q(N)$ is a real number, the vectors defined by $(6t(q)/\pi^2)\eta_q^* = B(q)(\gamma_q^* + \theta_q^*)/2$ and $(6t(q)/\pi^2)\kappa_q^* = B(q)(\gamma_q^* - \theta_q^*)/2$ where $B(q)$ is defined on $\mathcal{Q}$ by

$$
(19.11) \qquad \forall q = q_1 q_2^2 \in \mathcal{Q}, \quad B(q) = B(q_1 q_2^2) = q_1^{-2} q_2^{-1}
$$

are orthogonal. Of course $B(q)$ could be any positive quantity depending only on q. We have chosen it so as to minimize the error term in Lemma 19.6. See also Lemma 19.5. We further set

$$
(19.12) \qquad \varphi_q^* = \nabla_q \eta_q^*, \quad \psi_q^* = \nabla_q \kappa_q^*.
$$

We readily compute that

$$
(19.13) \qquad \begin{cases} \|\eta_q^*\|_q^2 = B(q)^2 (\phi(q) + c_q(N))/2, \\ \|\kappa_q^*\|_q^2 = B(q)^2 (\phi(q) - c_q(N))/2. \end{cases}
$$

Of course, when $q|N$, κ_q^* is the zero vector and so not of great interest. Otherwise here is a lemma that gives an *apriori* bound for their norms.

Lemma 19.3. *When $q \nmid N$, both norms $\|\eta_q^*\|_q^2$ and $\|\kappa_q^*\|_q^2$ lie between $B(q)^2\phi(q)/4$ and $B(q)^2\phi(q)$.*

Proof. Indeed $|c_q(N)| = \phi((N,q))$ divides strictly $\phi(q)$ and is hence at most $\phi(q)/2$. The lemma follows readily.   ◇◇◇

We now need to define the global scalar product. As an orthogonal system, we take the union of (φ_q^*) and of (ψ_q^*) but in the latter family, we remove the terms for which $q|N$.

Lemma 19.4. *Let q_1 and q_2 be too moduli, and q_3 their lcm. If u_{q_1} and v_{q_2} are respectively one of $\{\eta_{q_1}^*, \kappa_{q_1}^*\}$ and $\{\eta_{q_2}^*, \kappa_{q_2}^*\}$ then*

$$[\nabla_{q_1} u_{q_1}|\nabla_{q_2} v_{q_2}] = N[L_{q_3}^{\tilde{q}_1} u_{q_1}|L_{q_3}^{\tilde{q}_2} v_{q_2}]_{q_3} + \mathcal{O}\left(B(q_1)\sigma(q_1)B(q_2)\sigma(q_2)\right).$$

A lemma where we somehow used deliberately a complicated expression. As it turns out, if $q_1 \neq q_2$ or $u_{q_1} \neq v_{q_1}$, the local scalar product vanishes! To handle the hermitian properties of (φ_q^*) and (ψ_q^*), we could try something along the following lines:

$$[\varphi_q^*|\varphi_q^*] = [\nabla_q \eta_q^*|\nabla_q \eta_q^*] = [\Delta_q \nabla_q \eta_q^*|\eta_q^*]_q.$$

However $\Delta_q \nabla_q$ is *not* a multiple of the identity! However

$$\Delta_q \nabla_q h(c) = \left(\sum_{\substack{n \equiv c[q], \\ n \leq N}} 1\right) h(c)$$

so that we can $\Delta_q \nabla_q$ can be thought as a perturbation of the identity. We escaped from this complication in (16.5)-(16.10) by using a smooth majorant of the characteristic function of the interval $[1, N]$.

Proof. This expression is important in that it uses the structure of the Ramanujan sums. If we were to split the sum that defines the initial scalar product in classes modulo q_3, the remainder term would only be

$$\ll q_3 \sqrt{[u_{q_1}|u_{q_1}]_{q_1}[v_{q_2}|v_{q_2}]_{q_2}},$$

which looses a power of q_1 and one of q_2. The result we claim is obtained by appealing to $c_q(n) = \sum_{d|(n,q)} d\mu(q/d)$, an expression that expresses the fact that c_q is not an intricate function.   ◇◇◇

We are to majorize, when q is fixed:

(19.14) $$\sum_{q' \in \mathcal{Q}} |[\varphi_q^*|\varphi_{q'}^*]| + \sum_{q' \in \mathcal{Q}} |[\varphi_q^*|\psi_{q'}^*]|,$$

and

(19.15)
$$\sum_{q'\in\mathcal{Q}}||\psi_q^*|\varphi_{q'}^*|| + \sum_{q'\in\mathcal{Q}}||\psi_q^*|\psi_{q'}^*||.$$

Lemma 19.5. *When $Q_1, Q_2 \geq 100$, we have*

$$\sum_{\substack{q_1\leq Q_1 \\ q_2\leq Q_2 \\ (q_1,q_2)=1}} \frac{\mu^2(q_1)\sigma(q_1)}{q_1^2}\frac{\mu^2(q_2)q_2\sigma(q_2)}{q_2} \ll Q_2^2\operatorname{Log} Q_1.$$

We take

(19.16) $M_q(\varphi^*) = B(q)^2\dfrac{N(\phi(q)+c_q(N))}{2} + CB(q)\sigma(q)Q_2^2\operatorname{Log} Q$

and

(19.17) $M_q(\psi^*) = B(q)^2\dfrac{N(\phi(q)-c_q(N))}{2} + CB(q)\sigma(q)Q_2^2\operatorname{Log} Q$

for a C large enough that $M_q(\varphi^*)$ is more than (19.14) and that $M_q(\psi^*)$ is more than (19.15).

19.4. Removing the M_q's

We have chosen a somewhat intricate version of the M_q's to get rid of the bilinear form in $\omega_{i,j}$, but ultimately we will have to remove them to get smoother expressions.

Lemma 19.6. *If $|\beta_q| \leq B(q)^2(Nt(q)\phi(q)+\sqrt{Nq})^2$, then we have (for any $\varepsilon > 0$),*

$$\sum_{q\in\mathcal{Q}} M_q(\varphi_q^*)^{-1}\beta_q = \sum_{q\in\mathcal{Q}} \frac{2\beta_q}{NB(q)^2(\phi(q)+c_q(N))} + \mathcal{O}_\varepsilon(Q_2Q^{1+\varepsilon})$$

as well as

$$\sum_{\substack{q\in\mathcal{Q} \\ q\nmid N}} M_q(\psi_q^*)^{-1}\beta_q = \sum_{\substack{q\in\mathcal{Q} \\ q\nmid N}} \frac{2\beta_q}{NB(q)^2(\phi(q)-c_q(N))} + \mathcal{O}_\varepsilon(Q_2Q^{1+\varepsilon}).$$

Proof. A similar treatment applies to both expressions. For instance, the difference between the R.H.S. and the L.H.S. of the first one is, by appealing to Lemma 19.3, at most

$$Q_2^2Q^\varepsilon\sum_{q\in\mathcal{Q}} \frac{B(q)^2(N^2t(q)^2\phi(q)^2+Nq)B(q)\sigma(q)}{(N\phi(q)B(q)^2+B(q)\sigma(q)Q_2^2Q^\varepsilon)N\phi(q)B(q)^2}$$

which is bounded by

$$Q_2^2 Q^{\varepsilon'} \sum_{q \in \mathcal{Q}} \frac{Nt(q)^2 \phi(q) + 1}{NB(q) + Q_2^2} \leq Q_2^2 Q^{\varepsilon'} \sum_{q \in \mathcal{Q}} \left(\frac{Nt(q)^2 \phi(q)}{NB(q)} + \frac{1}{Q_2^2} \right)$$

for all $\varepsilon > 0$, where ε' tends to zero with ε. ◇◇◇

19.5. Approximating f and g

The first local approximation of $f(n) = \mu^2(n)$ is $\Delta_q^*(f)/q$. Lemma 19.1 ensures us that $N\gamma_q$ is an approximation of $\Delta_q(f)s$, which suggests that we take $\nabla_q \gamma_q$ (see (19.7)) as a local approximation to f, well more precisely $\varphi_q^* = \nabla_q \gamma_q^*$ since only the orthogonal projection of γ_q over $\mathfrak{M}(q)$ is of interest. This is the path we follow, and correspondingly, we approximate g by ψ. Before quantifying in a proper way this approximation, we need an *apriori* upper bound. In this part, the roles of f and g are completely similar and it is enough to handle the case of f. The statements are however complete.

Lemma 19.7. *We have for every $\varepsilon > 0$:*

$$\|[\varphi_q^*|f]\| + \|[\psi_q^*|f]\| \ll_\varepsilon B(q)|t(q)|\phi(q)N + B(q)\sqrt{N}qq^\varepsilon$$

and similarly by replacing f by g.

Proof. We compare $[\varphi_q^*|f]$ to $N[\eta_q^*|\gamma_q^*]_q$. First we note that $B(q)\gamma_q^* = (6t(q)/\pi^2)(\eta_q^* + \kappa_q^*)$, from which we deduce that

$$(19.18) \qquad [\eta_q^*|\gamma_q^*]_q = 3B(q)t(q)(\phi(q) + c_q(N))/\pi^2.$$

Next, we check that

$$[\varphi_q^*|f] = [\nabla_q \eta_q^*|f] = [\eta_q^*|\Delta_q f]_q = [\eta_q^*|N\gamma_q^*]_q + [\eta_q^*|\Delta_q(f) - N\gamma_q]_q.$$

Next Lemma 19.1 gives

$$[\eta_q^*|\Delta_q(f) - N\gamma_q]_q = \mathcal{O}_\varepsilon \left(B(q) \sum_{d|q} d\mu(q/d)\mathcal{O}(d^\varepsilon \sqrt{N/d}) \right)$$

from which we infer $\|[\varphi_q^*|f]\| \ll_\varepsilon B(q)(|t(q)|\phi(q)N + \sqrt{N}qq^\varepsilon)$ as claimed. ◇◇◇

Lemma 19.8. *Let (α_q) be complex numbers such that*

$$(19.19) \qquad B(q)^2 N\phi(q)|\alpha_q| \leq NB(q)|t(q)|\phi(q) + B(q)\sqrt{N}q.$$

We have that for every $\varepsilon > 0$

$$\sum_{q \in Q} \alpha_q [\varphi_q^* | f] = \sum_{q \in Q} \alpha_q [\varphi_q^* | \varphi_q^* + \psi_q^*] + \mathcal{O}_\varepsilon \left(Q^\varepsilon (\sqrt{N} + Q_2 Q_1) \right)$$

and similarly by replacing f by g and $\varphi^* + \psi^*$ by $\varphi^* - \psi^*$.

Proof. There comes

$$\left| \sum_{q \in Q} \alpha_q [\varphi_q^* | f - \varphi_q^* - \psi_q^*] \right| \le \sum_{q \in Q} \left| t(q) B(q)^2 \phi(q) \sum_{d|q} d\mu(q/d) \mathcal{O}(q^\varepsilon \sqrt{N/d}) \right|$$

$$\ll Q^\varepsilon (\sqrt{N} + Q_2 Q_1)$$

$\diamond\diamond\diamond$

Lemma 19.9. *Difference*

$$[f|f] - \sum_{q \in Q} M_q(\varphi^*)^{-1} |[\varphi_q^* | f]|^2 - \sum_{\substack{q \in Q \\ q \nmid N}} M_q(\psi^*)^{-1} |[\psi_q^* | f]|^2$$

is big-$\mathcal{O}$ of $Q^\varepsilon (N Q_1^{-2} + N Q_2^{-1} + Q_2 Q + \sqrt{N})$. The same bound holds when f is replaced with g.

Proof. As first step we approximate f by φ_q^* in the products $[\varphi_q^* | f]$ and $[\psi_q^* | f]$. To do so we set $\beta_q = M_q(\varphi^*)^{-1} [f | \varphi_q^*]$, whose modulus indeed verifies condition (19.19) up to a Q^ε, and write

$$\sum_{q \in Q} M_q(\varphi^*)^{-1} |[\varphi_q^* | f]|^2 = \sum_{q \in Q} \beta_q [\varphi_q^* | f].$$

By the preceding lemma, we can thus replace f by its local approximation, up to an admissible error term. We reiterate the process:

$$\sum_{q \in Q} M_q(\varphi^*)^{-1} |[\varphi_q^* | f]|^2 + \sum_{\substack{q \in Q \\ q \nmid N}} M_q(\psi^*)^{-1} |[\psi_q^* | f]|^2 =$$

$$\sum_{q \in Q} M_q(\varphi^*)^{-1} |[\varphi_q^* | \varphi_q^* + \psi_q^*]|^2 + \sum_{\substack{q \in Q \\ q \nmid N}} M_q(\psi^*)^{-1} |[\psi_q^* | \varphi_q^* + \psi_q^*]|^2$$

$$+ \mathcal{O}_\varepsilon \left(Q^\varepsilon (\sqrt{N} + Q_2 Q) \right).$$

Here we may replace $[\varphi_q^* | \varphi_q^* + \psi_q^*]$ and $[\psi_q^* | \varphi_q^* + \psi_q^*]$ respectively by $N[\eta_q^* | \gamma_q^*]_q$ and $N[\kappa_q^* | \gamma_q^*]_q$ up to a negligible error term. In the second step, we need to compute the R.H.S. First replace $M_q(\varphi^*)$ by $[\varphi_q^* | \varphi_q^*]$

with error term $\mathcal{O}(Q_2 Q^{1+\varepsilon})$ by Lemma 19.6, and then do the same with ψ^*. We check that

$$\big|[\eta_q^* | \gamma_q^*]_q\big|^2 \|\eta_q^*\|_q^{-2} + \big\|[\kappa_q^* | \gamma_q^*]_q\big|^2 \|\kappa_q^*\|_q^{-2} = \|\gamma_q^*\|_q^2.$$

Now $\|\gamma_q^*\|_q^2$ is computed in (19.10) and equals $\left(6/\pi^2\right)^2 \phi(q) t(q)^2$. We simply have to sum, complete the resulting series and estimate the resulting error term:

$$\sum_{q \in \mathcal{Q}} \phi(q) t(q)^2 = \sum_{\substack{q_1 \leq Q_1, q_2 \leq Q_2 \\ (q_1, q_2) = 1}} \frac{\mu^2(q_1 q_2) q_2}{\phi(q_1)\sigma(q_1)^2 \phi(q_2)\sigma(q_2)^2}$$

that is

$$\sum_{q \in \mathcal{Q}} \phi(q) t(q)^2 = \frac{\pi^2}{6} + \mathcal{O}(Q_1^{-2} + Q_2^{-1}).$$

Finally, recall that $[f|f] = (6/\pi^2)N + \mathcal{O}(\sqrt{N})$. The lemma follows readily.   ◇◇◇

19.6. Crossed products

Lemma 19.10. *The sum*

$$\sum_{q \in \mathcal{Q}} M_q(\varphi^*)^{-1} [f|\varphi_q^*][\varphi_q^*|g] + \sum_{\substack{q \in \mathcal{Q} \\ q \nmid N}} M_q(\psi^*)^{-1} [f|\psi_q^*][\psi_q^*|g]$$

equals, for every $\varepsilon > 0$:

$$N\mathfrak{S}(N) + \mathcal{O}_\varepsilon\left(Q^\varepsilon(\sqrt{N} + NQ_1^{-2} + NQ_2^{-1} + Q_2 Q)\right).$$

Proof.   We follow closely the proof of Lemma 19.9 and replace the quantity $\sum_{q \in \mathcal{Q}} M_q(\varphi^*)^{-1}[f|\varphi_q^*][\varphi_q^*|g]$ by $N\sum_{q \in \mathcal{Q}}[\gamma_q^*|\eta_q^*]_q[\eta_q^*|\theta_q^*]_q/\|\eta_q^*\|_q^2$ at the cost of an error term of size $\mathcal{O}_\varepsilon(Q^\varepsilon\sqrt{N} + Q_2 Q^{1+\varepsilon})$. A similar treatment for the part with ψ leads to $N\sum_{q \in \mathcal{Q}}[\gamma_q^*|\kappa_q^*]_q[\eta_q^*|\kappa_q^*]_q/\|\kappa_q^*\|_q^2$. We note here that

$$[\gamma_q^*|\eta_q^*]_q[\eta_q^*|\theta_q^*]_q\|\eta_q^*\|_q^{-2} + [\gamma_q^*|\kappa_q^*]_q[\kappa_q^*|\theta_q^*]_q\|\kappa_q^*\|_q^{-2} = [\gamma_q^*|\theta_q^*]_q.$$

Extension to a complete series costs $\mathcal{O}(N(Q_1^{-2} + Q_2^{-1}))$. The constant $\mathfrak{S}(N)$ appears here in the form

$$\mathfrak{S}(N) = \prod_p \left(1 - \frac{1}{p^2} + \frac{1 + c_p(N) + c_{p^2}(N)}{p^4}\right).$$

                                                                                  ◇◇◇

19.7. Main proof

Let us set

$$\langle h_1 | h_2 \rangle = \sum_{q \in \mathcal{Q}} M_q^{-1}(\varphi^*)[h_1|\varphi_q^*][\varphi_q^*|h_2] + \sum_{\substack{q \in \mathcal{Q} \\ q \nmid N}} M_q^{-1}(\psi^*)[h_1|\psi_q^*][\psi_q^*|h_2].$$

We use Cauchy's inequality on the semi[2] hermitian product $[f|g] - \langle f|g \rangle$ and get

$$(19.20) \qquad \left| [f|g] - \langle f|g \rangle \right| \leq \sqrt{\left([f|f] - \langle f|f \rangle \right) \cdot \left([g|g] - \langle g|g \rangle \right)},$$

which enables us to approximate $[f|g]$ by $\langle f|g \rangle$ which, in turn, we evaluate via Lemma 19.10. Total error term is

$$\mathcal{O}_\varepsilon \left(Q^\varepsilon (N Q_1^{-2} + N Q_2^{-1} + Q_2 Q + \sqrt{N}) \right).$$

Our choice of Q_1 and Q_2 yields the Theorem.

19.8. Afterthoughts

We insisted on taking care of both f and g while choosing our orthogonal system. However, we could have taken care of only one of them, since the other part will anyway not be of any use (as a kind of shorthand, we can say that only the orthornormal projection of g on $\mathbb{C} \cdot f$ has any effect. This is only shorthand because the involved hermitian product in the "orthonormal" above has not been specified). This would however have modified the error term since, in (19.20), only $[f|f] - \langle f|f \rangle$ would have been small.

Let us turn to a different consideration. In the proof of Lemma 19.9, we simply said "we may replace $[\varphi_q^*|\varphi_q^*]$ by $N\|\eta_q^*\|_q^2$" but the question arises to know why we did not do it at the very beginning! This would have required us to define a local model modulo q as a function over $\mathbb{Z}/q\mathbb{Z}$, which would have had drawbacks in other parts of the theory. A different approach looks promising: still use $\varphi_q^* = \nabla_q \gamma_q^*$ as a local model modulo q for f, but use $\tilde{\varphi}_q^*$ instead of φ_q^* in the definition of the scalar product, where $\Delta_q \tilde{\varphi}_q^* = \gamma_q^*$. In fact, most of our argument works because Δ_q and ∇_q are almost inverse of each other, so our choice is not so very wrong. It is indeed not wrong at all since we would have to compute $[\tilde{\varphi}_q^*|\tilde{\varphi}_{q'}^*]$. With our definition, we have very explicit expressions and are able to compute the corresponding scalar product.

---

[2]It may be not definite

19.9. Adding a prime and a squarefree number

The previous method would work with almost no changes to compute the number of representations of an integer N as a sum of a squarefree integer and a prime, with error term at most $\mathcal{O}_\varepsilon(N^{5/6+\varepsilon})$. We leave the details to the reader.

20 On a large sieve equality

This last chapter presents directions to investigate, some limitations, and other slightly off topic material. We use also this pretext to provide a simple introduction to some modern techniques. Let us finally point out that (Ramaré, 2007a) contains also material on this subject, but very different in nature. We omit it here.

20.1. Informal presentation

In our studies of additive problems, the main argument consists in showing that

$$\sum_q M_q(\varphi^*)^{-1}|[f|\varphi_q^*]|^2 \longrightarrow \|f\|_2^2.$$

The question that naturally arises is to determine which functions f will satisfy such a property, keeping in mind that we can choose φ_q^* in terms of f. We also want φ_q^* to be a $\nabla_q \eta_q^*$ (see (19.7)) for some function η_q^* from $\mathfrak{M}(q)$. This is not exactly what we did in chapter 10, where we multiplied a function $\nabla_q \eta_q^*$ by a function "with no arithmetic part" to get φ_q^*, but we ignore this aspect here.

In our scheme, we also have essentially $[f|\varphi_q^*] \simeq \|\varphi_q^*\|_2^2$ as well as $M_q(\varphi^*) \simeq \|\varphi_q^*\|_2^2$ so that what we really require is

$$\sum_q \|\varphi_q^*\|_2^2 \longrightarrow \|f\|_2^2.$$

Since $\Delta_q(f)$ defined in (4.20) is the best local model for f modulo q, the choice $\varphi_q^* = \nabla_q U_{\tilde{q} \to q} \Delta_q(f)$ is recommended, up to some rescaling. We have by (4.14)

$$U_{\tilde{q} \to q} \Delta_q(f)(n) = \sum_{a \bmod {}^* q} S_f(-a/q)e(na/q)$$

with $S_f(\alpha) = \sum_n f(n)e(n\alpha)$. Next, and since

$$[\nabla_q U_{\tilde{q} \to q} \Delta_q(f)|\nabla_q U_{\tilde{q} \to q} \Delta_q(f)] = \sum_n \left| \sum_{a \bmod {}^* q} S_f(a/q)e(na/q) \right|^2$$

$$= (N + \mathcal{O}(q^2)) \sum_{a \bmod {}^* q} |S_f(a/q)|^2,$$

we see that we should divide $U_{\tilde{q} \to q} \Delta_q(f)$ by $\sqrt{N}$, so that we can guess the functions we are looking for are the ones for which

(20.1) $$\sum_{q \in \mathcal{Q}} \sum_{a \bmod {}^* q} |S_f(a/q)|^2 \longrightarrow N \|f\|_2^2.$$

The symbol "$\longrightarrow$" is not exactly well defined, since several parameters may vary together, like the set $\mathcal{Q}$ of moduli and N, but somehow, one should have equality in the large sieve inequality up to a negligible error term. From here onwards, two courses of actions appear.

20.2. A detour towards limit periodicity

By a limit periodic set, we mean a set whose characteristic function is a uniform limit of linear combinations of periodical functions. Let us start with some generalities on such functions. We refrain from using $\|f\|_\infty$ to denote $\max_n |f(n)|$ where n ranges positive integers, because the notation $\|\cdot\|$ is already overloaded in this monograph.

20.2.1. Survey of the general theory.
As in most theories of almost periodicity, a central role is played by a kind of integral operator. Here the key will come from

(20.2) $$T_N(f, \alpha) = \sum_{1 \le n \le N} f(n)e(-n\alpha)/N$$

and we define $T_\infty(f, \alpha)$ when the sequence $(T_N(f, \alpha))_N$ converges as its limiting value. When f is periodic over $\mathbb{N}$, then $T_\infty(f, \alpha)$ indeed exists for all values of α and is 0 whenever $\alpha \notin \mathbb{Q}$. Furthermore the reader will readily check that

$$T_\infty(e(\cdot \alpha'), \alpha) = \begin{cases} 0 & \text{if } \alpha' \ne \alpha, \\ 1 & \text{if } \alpha' = \alpha. \end{cases}$$

Let us now consider a limit periodic function f, by which we mean a limit, according to the uniform norm on the positive integers, of a sequence of periodical functions. We claim that $T_\infty(f, \alpha)$ exists for every α and vanishes if $\alpha \notin \mathbb{Q}$.

Proof. Let α in $\mathbb{R}$ and let $\varepsilon > 0$. There is a periodic function g such that $\max_n |f(n) - g(n)| \le \varepsilon$. Thus, for every N, we have

$$|T_N(f, \alpha) - T_N(g, \alpha)| \le \varepsilon.$$

Since $(T_M(g, \alpha))_M$ is Cauchy, we find a $N_0(g, \varepsilon)$ such that, for $N, N' \ge N_0(g, \varepsilon)$, we have $|T_N(g, \alpha) - T_{N'}(g, \alpha)| \le \varepsilon$. Thus under the same condition, $|T_N(f, \alpha) - T_{N'}(f, \alpha)| \le 3\varepsilon$, meaning that the sequence $(T_N(f, \alpha))_N$

is also Cauchy. As a consequence, we can assert that this sequence indeed converges. Once this point is established, it is not difficult to see that $T_\infty(f, \alpha)$ vanishes when $\alpha \notin \mathbb{Q}$, a property inherited from the behaviour of periodical functions.   ◇◇◇

Now we have at our disposal a canonical approximation to limit periodic f by setting

$$(20.3) \qquad \Psi(f, q)(n) = \sum_{a \bmod q} T_\infty(f, a/q) e(na/q),$$

which happens to equal f when this function admits q as a period. Such an expression is convenient for our purpose and shows how contributions with $a \bmod^* q$ add up. But this is not the best one to show that it does indeed approximate f. To achieve this goal, note that the frequencies

$$(20.4) \qquad F(f; q, b) = \lim_{n \to \infty} \frac{q}{N} \sum_{\substack{n \le N, \\ n \equiv a[q]}} f(n)$$

are well-defined and that we also have

$$(20.5) \qquad \Psi(f, q)(n) = \sum_{b \bmod q} F(f; q, b) \mathbb{1}_{n \equiv b[q]}.$$

Next take $\varepsilon > 0$ and periodic g such that $\max_n |g(n) - f(n)| \le \varepsilon$. Let q be a period of g. We readily find that $|F(f; q, b) - F(g; q, b)| \le \varepsilon$ so that

$$\max_n |\Psi(f, q)(n) - \Psi(g; q)(n)| = \max_{b \bmod q} \max_{n \equiv b[q]} |\Psi(f, q)(n) - \Psi(g; q)(n)| \le \varepsilon.$$

We can take for q the sequence $\mathrm{lcm}_{d \le Q} d$ and the above to show that $(\Psi(f, q))_q$ converges uniformly towards f.

20.2.2. L^2-setting. If we select a limit periodic set $\mathcal{A}$ and a bound $N \ge 1$, the sequence $(\mathbb{1}_{n \in \mathcal{A}})_{n \le N}$ is the limit of $((\Psi(\mathbb{1}_\mathcal{A}, q)(n))_n)_q$. Such a sequence is thus a good candidate for (20.1). It is however unclear if the context of limit periodic functions is the proper one. Having almost periodic functions with *spectrum* in $\mathbb{Q}$ (the spectrum is the set of α such that $T_N(f, \alpha)$ does not vanish) is certainly helpful in constructing periodic approximation of f, and as such the context of Wiener or Marcinkiewicz spaces (see (Coquet *et al.*, 1977) and (Bertrandias, 1966)) appears to be relevant.

We check immediately that

$$|F(f; q, b)|^2 \le \liminf_{N \to \infty} \frac{q}{N} \sum_{\substack{n \le N, \\ n \equiv b[q]}} |f(n)|^2$$

from which we infer

$$(20.6) \qquad \sum_{a \bmod q} |T_\infty(f; a/q)|^2 \leq \liminf_{n \to \infty} \frac{1}{N} \sum_{n \leq N} |f(n)|^2.$$

However, the existence of the R.H.S. limit (as a limit and not as a lim inf) is far from obvious, though it is plausible. When $\lim_N \frac{1}{N} \sum_{n \leq N} |f(n)|^2$ indeed exists and is the limit of $\sum_{a \bmod q} |T_\infty(f; a/q)|^2$, then f is pseudo-periodic; in the context developed by (Bertrandias, 1966) and (Coquet *et al.*, 1977), it amounts to saying that the spectral measure associated to f is purely discrete. This statement is made with a fixed q.

A function f on integers is said to be $\mathfrak{B}^2$-almost periodic, i.e. almost periodic in the sense of Besicovitch, if there is a sequence of periodic functions f_q such that

$$\lim_{q \to \infty} \limsup_{N \to \infty} \frac{1}{N} \sum_{n \leq N} |f(n) - f_q(n)|^2 = 0.$$

The reader will find in (Schwartz & Spilker, 1994) the theory of such functions.

Brüdern went into similar considerations and cleared the situation further in (Brüdern, 2000-2004) by proving that

Theorem 20.1. *Let f be such that all $T_\infty(f; a/q)$ exist. Then we have equivalence between:*

(1) f is $\mathfrak{B}^2$-almost periodic,
(2) $\lim_N \frac{1}{N} \sum_{n \leq N} |f(n)|^2 = \sum_{q \geq 1} \sum_{a \bmod^* q} |T_\infty(f; a/q)|^2.$

As a consequence, he considered the problem of representing an integer as a sum of two elements from two sequences, one of which verifies inequality (20.6) as an equality, and both such that the averages $T_\infty(\mathbb{1}_\mathcal{A}; a/q)$ exist. The first step is the following corollary of the previous Theorem:

Corollary 20.1. *If f and g are such that all $T_\infty(g; a/q)$ and $T_\infty(f; a/q)$ exist, and moreover g is $\mathfrak{B}^2$-almost periodic, then*

$$\lim_N \frac{1}{N} \sum_{n \leq N} f(n)\overline{g(n)} = \sum_{q \geq 1} \sum_{a \bmod^* q} T_\infty(f; a/q)\overline{T_\infty(g; a/q)}.$$

From which he deduces for instance that there are infinitely many squarefree integers n such that $n + 1$ is also squarefree. This material was presented at several conferences but no published form exists as of now. He utilizes the circle method; our method clearly dispenses with it as it does in the case of squarefree numbers (see chapter 19), where we furthermore get a quantitative statement.

Schlage-Puchta went one step further in (Puchta, 2002), where the following Theorem is proved:

Theorem 20.2. *Let $\mathcal{N}$ be a set of integers and let f be its characteristic function. Then f is $\mathfrak{B}^2$-almost periodic if and only if the following three conditions are verified:*

(1) $\mathcal{N}$ has positive density.
(2) The frequencies $F(f; q, b)$ defined in (20.4) exist.
(3) We have

$$\sum_{q \geq 1} \sum_{a \bmod {}^* q} |T_\infty(f; a/q)|^2 = \lim_{n \to \infty} \frac{1}{N} \sum_{n \leq N} |f(n)|^2.$$

In particular, (Puchta, 2002) and (Brüdern, 2000-2004) prove that a set $\mathcal{N}$ with a multiplicative characteristic function and positive density satifies these conditions.

20.3. A large sieve equality: a pedestrian approach

We consider the problem of equality from a different angle and ask for a special form for f so as to satisfy (20.1). The form we choose is the one that appears in sieve theory, namely the convolution of a sequence with small support with the constant sequence $\mathbb{1}$.

We start with a simple result whose proof is illuminating.

Theorem 20.3. *Let $q \geq 2$ be an integer and let L and L_0 be two non-negative real numbers. For every arbitrary sequence of complex numbers $(b_m)_{m \leq M}$, we have*

$$\sum_{a \bmod {}^* q} \left| \sum_{\substack{L_0 < \ell \leq L_0 + L \\ m \leq M}} b_m e(a\ell m/q) \right|^2 - \phi(q) L^2 \left| \sum_{q|m} b_m \right|^2$$

$$\ll LB(\|b\|_2^2 Mq)^{1/2} \operatorname{Log} q + \|b\|_2^2 (Mq + q^2) \operatorname{Log}^2 q$$

with $\|b\|_2^2 = \sum_m |b_m|^2$ and $B = \sum_m |b_m|$.

Proof. We discuss according to whether $q|m$ or not, and sum over ℓ. Discarding the latter terms gives rise to the main term. To get a rigorous error term, take modulus and sum over a prime to q. We have

$$\sum_{\substack{L_0 < \ell \le L_0 + L \\ m \le M}} b_m e(a\ell m/q) = L^* \sum_{\substack{m \le M, \\ q|m}} b_m + \sum_{\substack{m \le M, \\ q \nmid m}} b_m \mathcal{O}(1/\|am/q\|)$$

where $\|\alpha\|$ stands for the distance to the nearest integer and L^* is the number of integer points in the interval $]L_0, L_0 + L]$. Summing over a ranging reduced residues classes, the error term is $\mathcal{O}$ of

$$L^* \sum_{\substack{m' \le M, \\ q|m'}} |b_{m'}| \sum_{\substack{m \le M, \\ q \nmid m}} |b_m| \sum_{a \bmod {}^* q} 1/\|am/q\|$$

$$+ \sum_{\substack{m,m' \le M, \\ q \nmid m, q \nmid m'}} |b_m||b_{m'}| \sum_{a \bmod {}^* q} 1/(\|am/q\|\|am'/q\|).$$

For the first one, proceed as follows: set $(m, q) = q/d < q$. Split summation over a according to classes modulo d; there are $\phi(q)/\phi(d) \le q/d$ elements per class where the latter inequality is proven by appealing to multiplicativity. Say $a \equiv b[d]$. We have

$$\sum_{c \bmod {}^* d} 1/\|cm/q\| \ll d(\operatorname{Log} d + 1) \ll d \operatorname{Log} q$$

which we multiply by q/d. This amounts to a contribution not more than

$$L \sum_{\substack{m' \le M, \\ q|m'}} |b_{m'}| \sum_{\substack{m \le M, \\ q \nmid m}} |b_m| \, \mathcal{O}(q \operatorname{Log} q) \ll LB\|b\|_2 \sqrt{M/q} \, q \operatorname{Log} q$$

which is $\mathcal{O}(LB\|b\|_2 (Mq)^{1/2} \operatorname{Log} q)$. It is a striking feature of this simple-minded proof that M/q occurs and not $M/q + 1$. As for the second part of the error term, we use $2|b_m b_{m'}| \le |b_m|^2 + |b_{m'}|^2$ to get it is not more – up to a multiplicative constant – than

$$\sum_{\substack{m \le M, \\ q \nmid m}} |b_m|^2 \sum_{a \bmod {}^* q} \sum_{\substack{m' \le M, \\ q \nmid m'}} 1/(\|am/q\|\|am'/q\|).$$

For the sum over m' we split the range of summation in interval of length q and get it is $\mathcal{O}((1 + M/q)q \operatorname{Log} q)$. We treat the sum over a as above and get a total contribution of

$$\mathcal{O}\big(\|b\|_2^2 (M + q)q \operatorname{Log}^2 q\big)$$

as required. The error due to the replacement of L^* with L is absorbed in the already existing error term. $\diamond \diamond \diamond$

The statement of the above Theorem can be simplified by using $B^2 \leq M\|b\|_2^2$, but this may lead to a severe loss when the sequence b has a small support. If this happens, a similar loss most probably occurs in the second part of the error term; the reader may try to recover this loss by inspecting the proof above: after using $2|b_m b_{m'}| \leq |b_m|^2 + |b_{m'}|^2$, we extend the summation over m' to every integers $\leq M$ and this can be costly (for instance when b_m is supported by the squares).

Summing over q, we get an impressive result which will compare easily with the theorem proved in section 20.5.

Corollary 20.2. *Let $\mathcal{Q}$ be a set of moduli, all $\leq Q$. For every sequence of complex numbers $(b_m)_{m \leq M}$, we have*

$$\sum_{q \in \mathcal{Q}} \sum_{a \bmod^* q} \left| \sum_{\substack{L_0 < \ell \leq L_0 + L \\ m \leq M}} b_m e(a\ell m/q) \right|^2 = L^2 \sum_{m,m' \leq M} b_m \overline{b_{m'}} (m, m')_{\mathcal{Q}}$$

$$+ \mathcal{O}\left(LB\|b\|_2 (MQ^3)^{1/2} \operatorname{Log} Q + \|b\|_2^2 (MLQ^{3/2} + MQ^2 + Q^3) \operatorname{Log}^2 Q \right)$$

with notations as in Theorem 20.3 and

$$(20.7) \qquad \forall m, m' \in \mathbb{N} \setminus \{0\}, \quad (m, m')_{\mathcal{Q}} = \sum_{\substack{t \in \mathcal{Q}, \\ t|m, t|m'}} \phi(t).$$

To understand the strength of this corollary, consider case $b_m = 1$ and $\mathcal{Q} = \{q \leq Q\}$. Then the main term is of size $(LM \operatorname{Log} M)^2$, while the error term is of size at most $M(MLQ^{3/2} + MQ^2 + Q^3) \operatorname{Log}^2 Q$ which is indeed an error term when $Q \leq L^{2/3}$. This is a considerable improvement on the large sieve inequality when M is relatively small! The latter would yield the upper bound $(L^2 M^2 + LMQ^2) \operatorname{Log}^2 M$ which is superseded by the above when $M \leq \min(L, Q^{1/2})$ and $LM \geq Q$; the most astonishing part of our result is that under some circumstances, we may take Q larger than $\sqrt{LM}$. For instance, with $M = N^\alpha$ and $L = N^{1-\alpha}$ for some $\alpha \in [0, 1/2]$, we can take $Q = N^{2(1-\alpha)/3}$, which is indeed larger than $\sqrt{N}$ if $\alpha \leq 1/4$.

We have taken here the convolution of $\mathbb{1}$ with (b_m), but we could easily replace $\mathbb{1}$ with any smooth function over this interval as we do in section 20.5. Since we do not require the Poisson summation formula here, we do not even need it to be differentiable at the endpoints of the interval of summation.

20.4. An application

The previous section contains results of a methodological character. As
such they have been presented in what we expect to be the simplest
setting, but applications call for slightly different statements. Let us
start with the following Lemma.

Lemma 20.1. *Let $q \geq 2$ be an integer and let N be a non-negative real
numbers. For every arbitrary sequence of complex numbers $(b_m)_{m \leq M}$,
we have*

$$\sum_{a \bmod^* q} \left| \sum_{\substack{\ell \geq 1, \\ m \leq M, \\ \ell m \leq N}} b_m e(a\ell m/q) \right|^2 - \phi(q) \left| \sum_{q|m} b_m [N/m] \right|^2$$

$$\ll NBq^{1/2} \operatorname{Log} q \sum_{\substack{m \leq M, \\ q|m}} \frac{|b_m|}{m} + \|b\|_2^2 (Mq + q^2) \operatorname{Log}^2 q$$

with $\|b\|_2^2 = \sum_m |b_m|^2$ and $B = \sum_m |b_m|$.

For the use we have in mind, namely the squarefree numbers, replacing the integer part $[N/m]$ by N/m up to a $\mathcal{O}(1)$ would be too costly
without any further assumptions on (b_m). We thus keep the main term
in this fairly raw format.

Proof.   The proof is simply an adaptation of the one given for Theorem 20.3. We start from

$$\Sigma = \sum_{\substack{\ell \geq 1, \\ m \leq M, \\ \ell m \leq N}} b_m e(a\ell m/q) = \sum_{\substack{m \leq M, \\ q|m}} b_m [N/m] + \sum_{\substack{m \leq M, \\ q \nmid m}} b_m \mathcal{O}(1/\|am/q\|)$$

Summing over a ranging reduced residues classes, we first get that

$$\Sigma = \phi(q) \left| \sum_{\substack{m \leq M, \\ q|m}} b_m [N/m] \right|^2$$

$$+ \mathcal{O}\left(\sum_{\substack{m \leq M, \\ q|m}} |b_m| \frac{N}{m} \sum_{a \bmod^* q} \sum_{\substack{m' \leq M, \\ q \nmid m'}} |b_{m'}|/\|am'/q\| \right.$$

$$\left. + \sum_{\substack{m, m' \leq M, \\ q \nmid m, q \nmid m'}} |b_m| |b_{m'}| \sum_{a \bmod^* q} 1/(\|am/q\| \|am'/q\|) \right)$$

where we handle the error term as in the proof of Theorem 20.3. ⋄⋄⋄

Summing over q, we infer the following Theorem.

Theorem 20.4. *Let $\mathcal{Q}$ be a set of moduli, all $\leq Q$. For every sequence of complex numbers $(b_m)_{m \leq M}$, we have*

$$\sum_{q \in \mathcal{Q}} \sum_{a \bmod {}^* q} \left| \sum_{\substack{\ell \geq 1, \\ m \leq M, \\ \ell m \leq N}} b_m e(a\ell m/q) \right|^2 = \sum_{m,m' \leq M} b_m \overline{b_{m'}} \sum_{\substack{q \in \mathcal{Q}, \\ q|(m,m')}} \phi(q) \left[\frac{N}{m}\right]\left[\frac{N}{m'}\right]$$

$$+ \mathcal{O}\big(NB\|b\|_2 \operatorname{Log}(MQ) + \|b\|_2^2 (MQ^2 + Q^3) \operatorname{Log}^2 Q\big)$$

with $\|b\|_2^2 = \sum_m |b_m|^2$ and $B = \sum_m |b_m|$.

Proof. We note that

$$\sum_{q \leq Q} \sum_{\substack{m \leq M, \\ q|m}} \frac{|b_m|}{m} \sqrt{q} \leq \sum_{m \leq M} \frac{|b_m|}{m} \sum_{q|m} \sqrt{q}$$

$$\leq \|b\|_2 \left(\sum_{m \leq M} \frac{1}{m^2} \Big(\sum_{q|m} \sqrt{q}\Big)^2 \right)^{1/2}.$$

and end the proof by noticing that

$$\sum_{m \leq M} \frac{1}{m^2} \Big(\sum_{q|m} \sqrt{q}\Big)^2 \ll \operatorname{Log} M$$

by appealing for instance to Theorem 21.1.   ⋄⋄⋄

When the main term in the above Theorem is of size about N^2 and $\|b\|_2^2$ is of size about M, the formula stated yields an asymptotic provided $Q, M = o(N^{2/3})$ and $QM = o(N)$.

Lemma 20.2. *For every sequence of bounded complex numbers $(c_d)_{d \leq D}$, we have*

$$\sum_{q \leq Q} \sum_{a \bmod {}^* q} \left| \sum_{\substack{\ell \geq 1, \\ d > D, \\ \ell d^2 \leq N}} c_d e(a\ell d^2/q) \right|^2 \ll NQ^2 D^{-2} + N^3 D^{-4} \operatorname{Log} N.$$

Proof. We simply expand the range of the inner summation over a to all of $\mathbb{Z}/q\mathbb{Z}$. Calling Σ the sum we want to estimate, this leads to

$$\Sigma \ll \sum_{D < d_1, d_2 \leq \sqrt{N}} |c_{d_1} c_{d_2}| \sum_{\substack{n_1, n_2 \leq N, \\ d_1^2 | n_1, \\ d_2^2 | n_2}} \sum_{\substack{q \leq Q, \\ q | n_1 - n_2}} q.$$

The diagonal terms $n_1 = n_2$ give rise to a contribution at most

$$NQ^2 \sum_{D < d_1, d_2 \leq \sqrt{N}} 1/[d_1^2, d_2^2] \ll NQ^2 \sum_{D < d_1, d_2 \leq \sqrt{N}} (d_1^2, d_2^2)/(d_1^2 d_2^2)$$

$$\ll NQ^2 \sum_\delta \phi(\delta) \left(\sum_{\substack{D < d \leq \sqrt{N}, \\ \delta | d}} 1/d^2 \right)^2$$

by using yet again Selberg's diagonalization process. When $\delta \leq D$, we bound the inner sum by $\mathcal{O}(1/(D\delta))$, while we bound it by $\mathcal{O}(1/\delta^2)$ otherwise. The total contribution of the diagonal terms is thus seen to be not more than $\mathcal{O}(NQ^2 D^{-2})$. Concerning the non-diagonal ones, we use

$$\sum_{q | m} q = m \prod_{p | m} (1 + 1/p) \leq m \cdot \exp\left(\sum_{p \leq m} 1/p \right) \ll m \operatorname{Log} m$$

to get a contribution of order at most:

$$\sum_{D < d_1 \leq d_2 \leq \sqrt{N}} \sum_{\substack{n_1, n_2 \leq N, \\ d_1^2 | n_1, \\ d_2^2 | n_2}} \frac{N}{d_1^2} \operatorname{Log} N \ll N^3 \operatorname{Log} N \sum_{D < d_1 \leq d_2 \leq \sqrt{N}} d_1^{-4} d_2^{-2}$$

$$\ll N^3 \operatorname{Log} N \sum_{D < d_1 \leq d_2 \leq \sqrt{N}} d_1^{-5} \ll N^3 D^{-4} \operatorname{Log} N$$

◇ ◇ ◇

Theorem 20.5. *For every $Q \leq N^{7/12 - 2\epsilon}$ with ϵ being positive and $\leq 1/6$, we have*

$$\sum_{q \leq Q} \sum_{a \bmod {}^* q} \left| \sum_{n \leq N} \mu^2(n) e(an/q) \right|^2 = (6/\pi^2) N^2 + \mathcal{O}(N^{2-\epsilon}).$$

We have not tried to get the best exponent instead of $7/12$, but have restrained our argument to remain somewhat general. This Theorem is of special interest: first, it offers a large sieve equality and second, we can even allow Q to be strictly larger than $N^{1/2}$. The reader will find in (Brüdern & Perelli, 1999) more information on the exponential sum over the squarefree numbers. It seems that the above Theorem is novel.

Proof.   We denote in this proof the constant $6/\pi^2$ by C to simplify the typographical work . We start as in section 19.2 with the formula $\mu^2(n) = \sum_{d^2|n} \mu(d)$ from which we infer

$$(20.8) \qquad \mu^2(n) = \sum_{\substack{d \leq D, \\ d^2|n}} \mu(d) + \sum_{\substack{d > D, \\ d^2|n}} \mu(d)$$

for some parameter $D \leq \min(N^{1/3}, Q)$. We then apply Theorem 20.4 with $m = d^2$, $M = D^2$ and $b_m = \mu(d)$ when $m = d^2$ and $b_m = 0$ otherwise. The main term reads

$$H = \sum_{q \leq Q} \phi(q) \left(\sum_{\substack{d \leq D, \\ q|d^2}} \mu(d)[N/d^2] \right)^2 .$$

When q is not cubefree, the inner summation vanishes, so we may write $q = q_1 q_2^2$ with q_1 and q_2 being squarefree and coprime. We set $q' = q_1 q_2$. The condition $q|d^2$ translates into $q'|d$. We have

$$\sum_{\substack{d \leq D, \\ q|d^2}} |\mu(d)[N/d^2]| \ll N/q'^2 \quad , \quad \sum_{\substack{d \leq D, \\ q|d^2}} |\mu(d)| \ll D/q',$$

so that

$$H = N^2 \sum_{q_1 q_2^2 \leq D^2} \mu^2(q_1 q_2)\phi(q_1)q_2\phi(q_2) \left(\sum_{\substack{d \leq D, \\ q_1 q_2|d}} \mu(d)/d^2 \right)^2 + \mathcal{O}(ND \operatorname{Log} D)$$

$$= C^2 N^2 \sum_{q_1 q_2^2 \leq D^2} \frac{\mu^2(q_1 q_2)\phi(q_1)\phi(q_2)}{q_1^4 q_2^3} \prod_{p|q_1 q_2} \left(1 - \frac{1}{p^2} \right)^{-2} + \mathcal{O}(N^2 D^{-1})$$

$$= CN^2 + \mathcal{O}(N^2 D^{-1})$$

where the last constant is a bit messy to compute: we first extend the summation to all $q_1 q_2$ with negligible error term and then proceed by multiplicativity. We first note that

$$\sum_{q_1, q_2 \geq 1} \frac{\mu^2(q_1 q_2)\phi(q_1)\phi(q_2)}{q_1^4 q_2^3} \prod_{p|q_1 q_2} \left(1 - \frac{1}{p^2} \right)^{-2}$$

$$= \sum_{q_1 \geq 1} \mu^2(q_1) \prod_{p|q_1} \frac{p-1}{(p^2-1)^2} \sum_{\substack{q_2 \geq 1, \\ (q_1, q_1)=1}} \mu^2(q_2) \prod_{p|q_1} \frac{p(p-1)}{(p^2-1)^2} = C'$$

say, and from then onward, continue routinely. We get

$$C' = \sum_{q_1 \geq 1} \mu^2(q_1) \prod_{p|q_1} \frac{p-1}{(p^2-1)^2 + p(p-1)} \prod_{p \geq 2} \left(1 + \frac{p(p-1)}{(p^2-1)^2}\right)$$

$$= \prod_{p \geq 2} \left(1 + \frac{p-1}{(p^2-1)^2 + p(p-1)}\right) \prod_{p \geq 2} \left(1 + \frac{p(p-1)}{(p^2-1)^2}\right)$$

$$= \prod_{p \geq 2} \frac{p^2}{p^2-1} = 1/C$$

as claimed. The error term in Theorem 20.4 is $\mathcal{O}((ND^{3/2} + D^3Q^2 + DQ^3)\operatorname{Log}^2(MQ))$. Let us define

$$\Sigma_1 = \sum_{q \leq Q} \sum_{a \bmod {}^*q} \left| \sum_{\substack{\ell \geq 1, \\ d \leq D, \\ \ell d^2 \leq N}} \mu(d)e(a\ell d^2/q) \right|^2$$

and Σ_2 with the size condition on d being reversed. We have just shown that $\Sigma_1 = CN^2 + \mathcal{O}(N^2D^{-1} + (ND^{3/2} + D^3Q^2 + DQ^3)\operatorname{Log}^2(MQ))$ while Lemma 20.2 yields the bound

$$\Sigma_2 \ll NQ^2D^{-2} + N^3D^{-4}\operatorname{Log} N.$$

Let us select $D = N^{1/4+\epsilon}$ and $Q = N^{7/12-2\epsilon}$. We readily get

$$\Sigma_2 \ll N^{2-2\epsilon}, \quad \Sigma_1 - CN^2 \ll N^{11/8+2\epsilon} + N^{23/12-\epsilon/2} + N^{2-4\epsilon} \ll N^{2-\epsilon}$$

since $\epsilon \leq 1/6$. We use

$$\sum_{q \leq Q} \sum_{a \bmod {}^*q} \left| \sum_{\substack{\ell \geq 1, \\ d \leq D, \\ \ell d^2 \leq N}} \mu(d)e(a\ell d^2/q) + \sum_{\substack{\ell \geq 1, \\ d > D, \\ \ell d^2 \leq N}} \mu(d)e(a\ell d^2/q) \right|^2$$

$$\leq \Sigma_1 - 2\Re \sum_{q \leq Q} \sum_{a \bmod {}^*q} \sum_{\substack{\ell \geq 1, \\ d \leq D \\ \ell d^2 \leq N}} \mu(d)e(a\ell d^2/q) \sum_{\substack{\ell \geq 1, \\ d > D \\ \ell d^2 \leq N}} \mu(d)e(-a\ell d^2/q) + \Sigma_2$$

and we invoke Cauchy inequality for the middle term to prove it is not more than $\sqrt{\Sigma_1 \Sigma_2} \ll N^{2-\epsilon}$; the Theorem is proved.   ◇◇◇

20.5. A large sieve equality: using more advanced technology

We get here an equality in the large sieve inequality in a wider range of M, in fact for M up to L. This time the range for Q will be restricted to be not more than the squareroot of the length of summation.

(Friedlander & Iwaniec, 1992) already considered the case of f being the convolution of $\mathbb{1}$ with a shortly supported arithmetical function and proved more refined estimates than ours. The proof below is essentially a simplified extract of theirs. Note however that in the Theorem below, we do not use the special set of moduli $\{q \leq Q\}$.

We take

$$(20.9) \qquad f(n) = \sum_{\substack{\ell m = n \\ m \leq M}} b_m g(\ell)$$

where g is smooth. More precisely, we assume that g is C^∞ and that

$$(20.10) \qquad |g^{(j)}(t)| \ll_j (\xi L)^{-j}$$

for all $j \geq 0$ for some parameter $\xi \in]1/L, 1]$ and $L \geq 1$. We further assume that $g(t) = 0$ if $t \geq 2L$. For simplicity, the reader may only consider the case $\xi = 1$ which will convey the main ideas and difficulties. The hypothesis on g is patterned on the following examples: take a C^∞ compactly supported function G on $[1, 2]$ and set $g(t) = G(t/L)$. Such a function will verify our assumptions with $\xi = 1$. In this way we can for instance approximate the characteristic function of the interval $[L, 2L]$. The parameter ξ is here to handle the precision of this approximation, and the smaller it is, the better the approximation. There are several examples to understand this point: first, we may simply take a function G en $[1, 2/\xi]$ and set $g(n) = G(n/(\xi L))$. Of course, $\xi = 1/L$ corresponds to the maximum precision. The example we took in section 1.2.1 does not concern a C^∞ function but is very closely related. Function b_ν defined in (15.6) has $|b_\nu^{(j)}(t)| \ll_j \delta^j$ for $j < 2\nu + 2$ (see (15.11)) and falls again in this category but for the assumption that it should vanish for $t \geq 2L$.

Theorem 20.6. *Let f be as above and $\mathcal{Q}$ be a set of moduli, all $\leq Q$. We have*

$$\sum_{q \in \mathcal{Q}} \sum_{a \bmod^* q} \left| \sum_n f(n) e(na/q) \right|^2 = \left(\int_{-\infty}^\infty g(w) dw \right)^2 \sum_{m, m'} b_m \overline{b_{m'}} (m, m')_{\mathcal{Q}}$$

$$+ \mathcal{O}_{\varepsilon, j} \left(L \|b\|_2^2 Q(Q + LM^2(M/(\xi L))^j)(LM)^\varepsilon \right)$$

for any $j \geq 2$ and any $\varepsilon > 0$. Notation $(m, m')_{\mathcal{Q}}$ is defined in (20.7).

In applications, $\xi L/M$ is at least a small power of LM, so, by taking j large enough, the term $LM^2(M/(\xi L))^j$ becomes not more than $(LM)^\varepsilon$. This is usually less than Q.

Proof. We only consider the case $\mathcal{Q} = \{q \leq Q\}$ for notational simplicity. We have

$$\Sigma(f,Q) = \sum_{q \leq Q} \sum_{a \bmod {}^* q} \left| \sum_{\substack{\ell m = n \\ m \leq M}} b_m g(\ell) e(\ell m a/q) \right|^2$$

$$= \sum_{m,m' \leq M} \sum_{\ell,\ell'} g(\ell)g(\ell') b_m \overline{b_{m'}} \sum_{q \leq Q} c_q(\ell m - \ell'm')$$

$$= \sum_{d \leq Q} dM(Q/d) \sum_{\substack{m,m',\ell,\ell' \\ m\ell \equiv m'\ell'[d]}} b_m \overline{b_{m'}} g(\ell)g(\ell')$$

$$= \sum_{d \leq Q} dM(Q/d) \sum_{|r| \leq N/d} \sum_{\substack{m,m',\ell,\ell' \\ m\ell - m'\ell' = dr}} b_m \overline{b_{m'}} g(\ell)g(\ell')$$

where $M(X) = \sum_{q \leq X} \mu(q)$ is the summatory function of the Moebius function.

Next, we have $m\ell - m'\ell' = dr$ and thus $\ell m \equiv dr[m']$. Let $(m,m') = \delta$, a divisor of dr. We set $m = \delta n$, $m' = \delta n'$ and $k = dr/\delta$. We define

$$(20.11) \qquad S_{dr} = \sum_{\delta | dr} \sum_{(m,m')=\delta} b_m \overline{b_{m'}} \sum_{\ell \equiv \overline{n}k[n']} g(\ell)g\left(\frac{n\ell - k}{n'}\right)$$

and get

$$S_{dr} = \sum_{\delta | dr} \sum_{(m,m')=\delta} \frac{b_m \overline{b_{m'}}}{n'} \sum_{h \in \mathbb{Z}} e\left(-\frac{\overline{n}kh}{n'}\right) \int_{-\infty}^{\infty} g(u)g\left(\frac{nu-k}{n'}\right) e\left(\frac{uh}{n'}\right) du$$

$$= \sum_{\delta | dr} \sum_{(m,m')=\delta} \frac{b_m \overline{b_{m'}}}{nn'} \sum_{h \in \mathbb{Z}} e\left(-\frac{\overline{n}kh}{n'}\right) \int_{-\infty}^{\infty} g\left(\frac{u}{n}\right) g\left(\frac{u-k}{n'}\right) e\left(\frac{uh}{nn'}\right) du$$

by Poisson summation formula. In this expression, we shall separate h into three ranges: $h = 0$ gives the main term, $0 < |h| < H$ could be treated in a non trivial way, something we do not wish to dwelve on here, while the contribution with $|h| \geq H$ will be discarded simply because Fourier coefficients tend to zero when the argument tends to infinity.

20.5.1. $h = 0$. The corresponding part of $\Sigma(f, Q)$ reads

$$\Sigma_0(f, Q) = \sum_{d \leq Q} dM\left(\frac{Q}{d}\right) \sum_{\substack{|r| \leq N/d,\ (m, m') = \delta \\ \delta | dr}} \sum \frac{b_m b_{m'}}{nn'} \int_{-\infty}^{\infty} g\left(\frac{u}{n}\right) g\left(\frac{u-k}{n'}\right) du$$

Here, δ being fixed, we want to sum over r. We need to have $\delta/(\delta, d)|r$ so that with $r = s\delta/(\delta, d)$

$$\sum_{|s| \leq N\frac{(d,\delta)}{d\delta}} g\left(\frac{u - ds/(\delta, d)}{n'}\right) = \int_{-\infty}^{\infty} g\left(\frac{u - vd/(\delta, d)}{n'}\right) dv + \mathcal{O}(1)$$

$$= \frac{(\delta, d)n'}{d} \int_{-\infty}^{\infty} g(w) dw + \mathcal{O}(1),$$

from which we infer

$$(20.12) \quad \sum_{\substack{|r| \leq N/d \\ \delta | dr}} \frac{1}{nn'} \int_{-\infty}^{\infty} g\left(\frac{u}{n}\right) g\left(\frac{u-k}{n'}\right) du$$

$$= \frac{(\delta, d)}{d} \left(\int_{-\infty}^{\infty} g(w) dw\right)^2 + \mathcal{O}(L/n').$$

The change of variable $u - k = v$ enables us to exchange roles of n and n', resulting in an error term of $\mathcal{O}(L/(n + n'))$.

Treatment of the main term when $h = 0$. Plugging such an estimate back into $\Sigma_0(f, Q)$, we get, for the main term

$$\left(\int_{-\infty}^{\infty} g(w) dw\right)^2 \sum_{m, m'} b_m \overline{b_{m'}} \sum_{d \leq Q} ((m, m'), d) M(Q/d).$$

Separate (m, m') and d by appealing to $\ell = \sum_{t | \ell} \phi(t)$ and get that the above is

$$\left(\int_{-\infty}^{\infty} g(w) dw\right)^2 \sum_{m, m'} b_m \overline{b_{m'}} \sum_{t | (m, m')} \phi(t) \sum_{\substack{d \leq Q \\ t | d}} M(Q/d).$$

We check that $\sum_{t | d \leq Q} M(Q/d) = 1$ if $t \leq Q$, and 0 otherwise. The main term now reads

$$\left(\int_{-\infty}^{\infty} g(w) dw\right)^2 \sum_{m, m'} b_m \overline{b_{m'}} (m, m').$$

Treatment of the error term when $h = 0$. Plugging (20.12) into the definition of $\Sigma_0(f, Q)$, we get the error term

$$QL \sum_{d \leq Q} \sum_{(m,m')=\delta} \frac{b_m \overline{b_{m'}}}{n + n'} \ll Q^2 L \sum_{m,m'} (m, m') \frac{|b_m \overline{b_{m'}}|}{m + m'}$$

We separate m and m' in two steps by using $\ell = \sum_{t|\ell} \phi(t)$ and then noting $2|b_m \overline{b_{m'}}| \leq |b_m|^2 + |b_{m'}|^2$. We get the above to be not more than

$$Q^2 L \sum_t \phi(t) \sum_{t|m} |b_m|^2 \sum_{t|m'} \frac{1}{m + m'} \ll_\varepsilon Q^2 L M^\varepsilon \|b\|_2^2.$$

20.5.2. $|h| \geq H$. We simply use a bound for the Fourier coefficient that is obtained by integrating $j \geq 2$ times.

$$\left| \int_{-\infty}^{\infty} g(v/n) g\left(\frac{v - k}{n'}\right) e\left(\frac{vh}{nn'}\right) dv \right| \ll \left(\frac{nn'}{h}\right)^j \left(\frac{1}{n} + \frac{1}{n'}\right)^j (\xi L)^{-j} L$$

$$\ll \left(\frac{n + n'}{h \xi L}\right)^j L \ll \left(\frac{M}{h \xi L \delta}\right)^j L$$

so the contribution to $\Sigma(f, Q)$ is at most

$$\sum_{d \leq Q} d(Q/d) \sum_{|r| \leq N/d} \sum_{|h| \geq H} \sum_{(m,m')=\delta|dr} |b_m b_{m'}| \left(\frac{M}{h \xi L \delta}\right)^j L$$

$$\ll Q \frac{M^2 L}{\xi} \|b\|_2^2 \left(\frac{M}{\xi L H}\right)^{j-1} \text{Log}(LM)$$

because the summation over h converges by the assumption $j \geq 2$ and since $N \ll LM$. As it turns out, our statement corresponds to $H = 1$.

20.5.3. $0 < |h| < H$. This subpart has no reasons to be, since we take $H = 1$. It would have become necessary to handle this case if we were to take $\xi L < M$.

$$\diamond \diamond \diamond$$

20.6. Equality in the large sieve inequality, II

We should compare the main term in Theorems 20.3 and 20.6 to $N \|f\|_2^2$ where N is supposedly the length, a notion that is not clearly defined here. In the large sieve inequality, N is an *upper bound* for the length. What is clear is that N should be of order LM. We consider only the simpler case of Theorem 20.3. Let us express $\|f\|_2^2$ in another manner:

$$\|f\|_2^2 = \sum_{m,m'\leq M} b_m\overline{b_{m'}} \sum_{\substack{\ell,\ell'\leq L, \\ \ell m=\ell'm'}} 1$$

In the inner sum, we write $\delta = (m, m')$ and $m = \delta n$ as well as $m' = \delta n'$. We should have $\ell = nh$ and $\ell' = n'h$ and we get

$$\|f\|_2^2 = L \sum_{m,m'\leq M} \frac{b_m\overline{b_{m'}}}{\max(m,m')}(m,m') + \mathcal{O}(M\|b\|_2^2).$$

This is to be compared with the main term we got on taking $\mathcal{Q} = \{q \leq Q\}$ and $Q \geq M$, namely:

$$(L/M) \sum_{m,m'\leq M} b_m\overline{b_{m'}}(m,m').$$

Both expressions are close but not close enough and it is likely that the latter should be a fraction of the former. The case $b_m = 1$ when $m \in]M/2, M]$, and 0 otherwise is enlightening. We see directly that

$$\sum_{M/2<m,m'\leq M} \frac{(m,m')}{\max(m,m')} = \sum_{d\leq M} \phi(d)\ 2 \sum_{\substack{M/2<m\leq m'\leq M, \\ d|m,d|m'}} \frac{1}{m'} + \mathcal{O}(M)$$

which we readily evaluate. It is

$$2\sum_{d\leq M} \frac{\phi(d)}{d} \sum_{\substack{M/2<m\leq M, \\ d|m}} \left(\text{Log}\,\frac{M}{m} + \mathcal{O}\left(\frac{d}{m}\right)\right) + \mathcal{O}(M)$$

$$= (1 - \text{Log}\,2)M \sum_{d\leq M} \frac{\phi(d)}{d^2} + \mathcal{O}(M)$$

$$= (1 - \text{Log}\,2)M \cdot C\,\text{Log}\,M + \mathcal{O}(M)$$

for a positive constant C while one readily checks that

$$(1/M) \sum_{M/2<m,m'\leq M} (m,m') = \frac{M}{4} \cdot C\,\text{Log}\,M + \mathcal{O}(M).$$

This example shows that a loss of a multiplicative constant is to be expected. We are really interested in what happens when one takes sieve weights, in which case b_m varies in sign while ℓ is not constrained as in Theorem 20.1, but I expect a similar phenomenom to happen. It is however out of the scope of this monograph.

20.7. The large sieve inequality reversed

(Duke & Iwaniec, 1992) proved a very interesting reversed large sieve
inequality that we only state here. We are somehow off topic.

Theorem 20.7. *Let $(b_n)_{n \geq 1}$ be a sequence of complex numbers. For
$M \geq 2N \geq 4$, there exist $Q \leq \sqrt{N}$ and a smooth function f supported
on a subinterval of $[M - N, M + 2N]$ of length $Y = Q\sqrt{N}$ and whose
derivatives verify $|f^{(j)}| \ll_j (\operatorname{Log} Y)/Y^j$ for any $j \geq 0$, such that*

$$\sum_{M < n \leq M+N} |b_n|^2 \leq \sum_{q \leq Q} (qQ)^{-1} \sum_{a \bmod q} \left| \sum_n b_n f(n) e(na/q) \right|^2.$$

The reader should be wary of the apparently small changes in the
quantities considered: first, the summation runs over all a's modulo q
and not only over the invertible residue classes ans secondly, we divide
by $1/q$ and *not* by $1/Q$. This last change has momentous consequences
which are better described by looking at the case $b_n = 1$. The right-hand
side of the above equation is then of order $NQ \operatorname{Log} Q$ while the left-hand
side is only of size N! We gather by inspecting this example that the
above inequality should be used only when $\sum_n b_n e(na/q)$ is expected to
be negligible for all small q's.

21 Appendix

21.1. A general mean value estimate

Here is a theorem inspired by (Halberstam & Richert, 1971) but where we take care of the values of our multiplicative function on powers of primes as well. The reader will find in (Martin, 2002) an appendix with a similar result. Moreover, we present a completely explicit estimate, which complicates the proof somewhat. In (Cazaran & Moree, 1999), the reader will find, inter alia, a presentation of many results in the area, a somewhat different exposition as well as a modified proof: the authors achieve there a better treatment of the error term by appealing to a preliminary sieving.

Theorem 21.1. *Let g be a non-negative multiplicative function. Let κ, L and A be three non-negative real parameters such that*

$$
\begin{cases}
\sum_{\substack{p \geq 2, \nu \geq 1 \\ w < p^\nu \leq Q}} g(p^\nu) \operatorname{Log}(p^\nu) = \kappa \operatorname{Log} \dfrac{Q}{w} + \mathcal{O}^*(L) & (Q > w \geq 1), \\[2em]
\displaystyle\sum_{p \geq 2} \sum_{\nu, k \geq 1} g(p^k) g(p^\nu) \operatorname{Log}(p^\nu) \leq A.
\end{cases}
$$

Then, when $D \geq \exp(2(L + A))$, we have

$$
\sum_{d \leq D} g(d) = C \left(\operatorname{Log} D\right)^\kappa \left(1 + \mathcal{O}^*(B / \operatorname{Log} D)\right)
$$

with

$$
\begin{cases}
C = \dfrac{1}{\Gamma(\kappa + 1)} \prod_{p \geq 2} \left\{ \left(1 - \dfrac{1}{p}\right)^\kappa \sum_{\nu \geq 0} g(p^\nu) \right\}, \\[1.5em]
B = 2(L + A)\left(1 + 2(\kappa + 1)e^{\kappa + 1}\right).
\end{cases}
$$

If in many applications the dependence in L is important, the one in A is most often irrelevant. In the context of the sieve, κ is called the *dimension* of the sieve: it is the parameter that determines the size of the average we are to compute and is, of course, of foremost importance. Let us mention in this direction that (Rawsthorne, 1982) obtains a one-sided result from one-sided hypothesis, following a path already thread in (Iwaniec, 1980).

Proof. Let us start with the idea of (Levin & Fainleib, 1967):

$$G(D) \operatorname{Log} D = \sum_{d \le D} g(d) \operatorname{Log} \frac{D}{d} + \sum_{d \le D} g(d) \operatorname{Log} d$$

$$= \sum_{d \le D} g(d) \operatorname{Log} \frac{D}{d} + \sum_{\substack{p \ge 2, \nu \ge 1 \\ p^\nu \le D}} g(p^\nu) \operatorname{Log}(p^\nu) \sum_{\substack{\ell \le D/p^\nu \\ (\ell, p) = 1}} g(\ell).$$

Next we set

$$(21.1) \quad \begin{cases} G_p(X) = \displaystyle\sum_{\substack{\ell \le X \\ (\ell, p) = 1}} g(\ell) \\[4mm] T(D) = \displaystyle\sum_{d \le D} g(d) \operatorname{Log} \frac{D}{d} = \int_1^D G(t) \frac{dt}{t}, \end{cases}$$

so that we can rewrite the above as

$$G(D) \operatorname{Log}(D) = T(D) + \sum_{\substack{p \ge 2, \nu \ge 1 \\ p^\nu \le D}} g(p^\nu) \operatorname{Log}(p^\nu) G_p(D/p^\nu).$$

Moreover

$$G_p(X) = G(X) - \sum_{k \ge 1} g(p^k) G_p(X/p^k)$$

which, when combined with our hypothesis, yields

$$G(D) \operatorname{Log}(D) = T(D) + \sum_{\substack{p \ge 2, \nu \ge 1 \\ p^\nu \le D}} g(p^\nu) \operatorname{Log}(p^\nu) G(D/p^\nu) + \mathcal{O}^*(A G(D))$$

$$= T(D) + \sum_{d \le D} g(d) \sum_{\substack{p \ge 2, \nu \ge 1 \\ p^\nu \le D/d}} g(p^\nu) \operatorname{Log}(p^\nu) + \mathcal{O}^*(A G(D))$$

$$= T(D)(\kappa + 1) + \mathcal{O}^*((L + A) G(D))$$

which we rewrite as

$$(\kappa + 1) T(D) = G(D) \operatorname{Log} D \ (1 + r(D))$$

$$\text{with } r(D) = \mathcal{O}^* \left(\frac{L + A}{\operatorname{Log} D} \right).$$

We see the previous equation as a differential equation. We set

$$\exp E(D) = \frac{(\kappa + 1) T(D)}{(\operatorname{Log} D)^{\kappa+1}} = \frac{G(D)}{(\operatorname{Log} D)^\kappa} (1 + r(D))$$

getting for $D \geq D_0 = \exp(2(L+A))$

$$E'(D) = \frac{T'(D)}{T(D)} - \frac{(\kappa + 1)}{D \operatorname{Log} D} = \frac{-r(D)(\kappa + 1)}{(1 + r(D))D \operatorname{Log} D}$$

$$= \mathcal{O}^* \left(\frac{2(L+A)(\kappa + 1)}{D(\operatorname{Log} D)^2} \right)$$

since $|r(D)| \leq 1/2$ when $D \geq D_0$ and on computing $T'(D)$ through (21.1). Now, still for $D \geq D_0$, we have

$$E(\infty) - E(D) = \int_D^\infty E'(t)dt = \mathcal{O}^* \left(\frac{2(L+A)(\kappa + 1)}{\operatorname{Log} D} \right).$$

Gathering our results, and using $\exp(x) \leq 1 + x \exp(x)$ valid for $x \geq 0$, we infer that

$$\frac{G(D)}{(\operatorname{Log} D)^\kappa} = \frac{\exp E(D)}{1 + r(D)} = \frac{e^{E(\infty)}}{1 + r(D)} \left(1 + \mathcal{O}^* \left(\frac{2(L+A)}{\operatorname{Log} D} (\kappa + 1) e^{\kappa + 1} \right) \right).$$

We next use $1/(1+x) \leq 1 + 2x$ valid when $0 \leq x \leq \frac{1}{2}$ and $(1+x)(1+y) \leq (1 + 2x + y)$ valid for $x, y \geq 0$ and $y \leq 1$ to infer

$$\frac{G(D)}{(\operatorname{Log} D)^\kappa} = e^{E(\infty)} \left(1 + \mathcal{O}^* \left(\frac{2(L+A)}{\operatorname{Log} D} (1 + 2(\kappa + 1) e^{\kappa + 1}) \right) \right).$$

This ends the main part of the proof. We are to identify $e^{E(\infty)} = C$. Note that the above proof is *apriori* wrong since $T'(D) \neq G(D)/D$ at the discontinuity points of G, but we simply have to restrict our attention to non integer D's and then proceed by continuity.

An expression for C. We define, for s a positive real number,

$$D(g, s) = \sum_{d \geq 1} \frac{g(d)}{d^s} = s \int_1^\infty G(D) \frac{dD}{D^{s+1}}$$

$$= sC \int_1^\infty (\operatorname{Log} D)^\kappa \frac{dD}{D^{s+1}} + \mathcal{O} \left(sC \int_1^\infty (\operatorname{Log} D)^{\kappa - 1} \frac{dD}{D^{s+1}} \right)$$

$$= C \left(s^{-\kappa} \Gamma(\kappa + 1) + \mathcal{O}(s^{1-\kappa} \Gamma(\kappa)) \right)$$

and consequently

$$C = \lim_{s \to 0^+} D(g, s) s^\kappa \Gamma(\kappa + 1)^{-1}$$

$$= \lim_{s \to 0^+} D(g, s) \zeta(s + 1)^{-\kappa} \Gamma(\kappa + 1)^{-1}.$$

It is then fairly easy to check that the Eulerian product

$$\prod_{p \geq 2} \left\{ \left(1 - \frac{1}{p} \right)^\kappa \sum_{\nu \geq 0} g(p^\nu) \right\}$$

is convergent with value $C\Gamma(\kappa + 1)$ as required.   ◇◇◇

21.2. A first consequence

It is not difficult by following (Wirsing, 1961) to derive a stronger mean value result from Theorem 21.1. Since it will be required in one of the applications below, and since all the necessary material has been already exposed, we include one such result.

Theorem 21.2. *Let f be a non-negative multiplicative function and κ be non-negative real parameter such that*

$$
\begin{cases}
\displaystyle\sum_{\substack{p\geq 2,\nu\geq 1\\ p^\nu\leq Q}} f(p^\nu)\,\mathrm{Log}(p^\nu) = \kappa Q + \mathcal{O}(Q/\mathrm{Log}(2Q)) & (Q\geq 1),\\[2em]
\displaystyle\sum_{p\geq 2}\sum_{\substack{\nu,k\geq 1,\\ p^{\nu+k}\leq Q}} f(p^k)f(p^\nu)\,\mathrm{Log}(p^\nu) \ll \sqrt{Q},
\end{cases}
$$

then we have

$$
\sum_{d\leq D} f(d) = \kappa\, C\cdot D\,(\mathrm{Log}\,D)^{\kappa-1}\,(1+o(1))
$$

where C is as in Theorem 21.1.

Proof. We proceed as in Theorem 21.1. Write

$$
S(D) = \sum_{d\leq D} f(d).
$$

By using Theorem 9.2 followed by an application of Theorem 21.1, we readily obtain the following apriori bound

(21.2) $$ S(D) \ll D(\mathrm{Log}(2D))^{\kappa-1}. $$

Consider now $S^*(D) = \sum_{d\leq D} f(d)\,\mathrm{Log}\,d$. Proceeding as in the proof of Theorem 21.1, we get

$$
S^*(D) = \sum_{\substack{p\geq 2,\nu\geq 1\\ p^\nu\leq D}} f(p^\nu)\,\mathrm{Log}(p^\nu) \sum_{\substack{\ell\leq D/p^\nu\\ (\ell,p)=1}} f(\ell)
$$

$$
= \sum_{\ell\leq D} f(\ell) \sum_{\substack{p\geq 2,\nu\geq 1\\ p^\nu\leq D/\ell,\\ (p,\ell)=1}} f(p^\nu)\,\mathrm{Log}(p^\nu)
$$

so that $S^*(D)$ equals

$$
\sum_{\ell\leq D} f(\ell) \sum_{\substack{p\geq 2,\nu\geq 1\\ p^\nu\leq D/\ell}} f(p^\nu)\,\mathrm{Log}(p^\nu) - \sum_{\ell\leq D} f(\ell) \sum_{\substack{p\geq 2,\nu,k\geq 1\\ p^{\nu+k}\leq D/\ell,\\ (p,\ell)=1}} f(p^\nu)f(p^k)\,\mathrm{Log}(p^\nu)
$$

We use our hypothesis on this expression and conclude that

$$S^*(D) = \kappa D \sum_{\ell \le D} f(\ell)/\ell + \mathcal{O}\left(Q \sum_{\ell \le D} \frac{f(\ell)}{\ell \operatorname{Log}(2Q/\ell)}\right) + \mathcal{O}\left(\sqrt{Q} \sum_{\ell \le D} \frac{f(\ell)}{\sqrt{\ell}}\right).$$

Both error terms are shown to be $\mathcal{O}(Q \operatorname{Log}(2Q)^{\kappa-1})$ by appealing to (21.2) while the main term is evaluated via Theorem 21.1. We finally use an integration by parts:

$$S(D) = 1 + \int_2^D S^*(t) \frac{dt}{t \operatorname{Log}^2 t} + \frac{S^*(D)}{\operatorname{Log} D}$$

to get the claimed asymptotic.   ◇◇◇

21.3. Some classical sieve bounds

Using Corollary 2.1 with Theorem 21.1 yields some classical sieve bounds.

Sums of two squares. Let us recall that a positive integer is a sum of two *coprime* squares if and only if its prime factor decomposition contains only powers of 2 or of primes congruent to 1 modulo 4. Let us call $\mathcal{B}$ the set of such numbers.

We consider the compact set $\mathcal{K}$ built as follows: if p is $= 2$ or a prime $\equiv 1[4]$, $\mathcal{K}_p$ is $\mathbb{Z}/p\mathbb{Z}$, and if $p \equiv 3[4]$, $\mathcal{K}_p = \mathcal{U}_p$. We then build $\mathcal{K}_d$ for squarefree d by split multiplicativity and, in general, $\mathcal{K}_d$ by lifting $\mathcal{K}_\ell$ in a trivial way through $\sigma_{d \to \ell}$ (see (2.1)), where ℓ is the squarefree kernel of d. The resulting compact set is multiplicatively split and squarefree. We readily check that

$$\begin{cases} h(2^\nu) = h(p^\nu) = 0 & \text{when } p \equiv 1[4] \text{ and } \nu \ge 1, \\ h(p) = 1/(p-1) \text{ and } h(p^\nu) = 0 & \text{when } p \equiv 3[4] \text{ and } \nu \ge 2. \end{cases}$$

Let $b(q)$ be the characteristic function of the integers whose prime factors are all $\equiv 3[4]$. We find that

$$(21.3) \qquad\qquad G_1(Q) = \sum_{q \le Q} \frac{\mu^2(q)b(q)}{\phi(q)}$$

to which we apply Theorem 21.1 with $\kappa = 1/2$. We get

$$(21.4) \qquad\qquad G_1(Q) \sim B\sqrt{\operatorname{Log} Q}$$

with B the product over all primes of $\sqrt{1 - p^{-1}}$ when $p \equiv 1[4]$ and $1/\sqrt{1 - p^{-1}}$ when $p \equiv 3[4]$, which product we multiply by $\sqrt{2/\pi}$ (the contribution of the factor 2 and of the Γ-factor, since $\Gamma(1/2) = \sqrt{\pi}$). On taking $Q = \sqrt{N}/\operatorname{Log} N$, we find that the number of elements in $\mathcal{B}$ in an interval of length N is not more than

(21.5) $(1 + o(1))\frac{\sqrt{2}}{B} N/\sqrt{\operatorname{Log} N}$.

This is to be compared to the number of elements of this sequence in the initial interval $[1, N]$. Using Theorem 21.2, we find that this cardinality is

(21.6) $\left| \{ b \in \mathcal{B}, b \leq N \} \right| = (1 + o(1)) \dfrac{\sqrt{2}}{\pi} \times \dfrac{\sqrt{2}}{B} N / \sqrt{\operatorname{Log} N}$.

so that our upper bound is about $\pi/\sqrt{2} = 2.22\ldots$ times off the exact answer in this case. The combinatorial sieve is able to get the asymptotic here, or even in the case of an interval $[M + 1, M + N]$, when N is not too large with respect to M.

On the number of prime twins. We will give an upper bound for the number of prime twins up to N, as N goes to infinity, by applying Selberg sieve. We already gave such a bound in chapter 9 by using our local models. The compact set we take is simply $\mathcal{K} = \mathcal{U} \cap (\mathcal{U} - 2)$ as was the case then. It is multiplicatively split as well as squarefree. For the associated function h, we readily find that

$$\begin{cases} h(2) = 1 \quad \text{and} \quad h(2^\nu) = 0 \qquad \text{if } \nu \geq 2, \\ h(p) = 2/(p - 2) \quad \text{and} \quad h(p^\nu) = 0 \text{ if } p \geq 3 \text{ and } \nu \geq 2. \end{cases}$$

This gives us

(21.7) $G_1(Q) = \displaystyle\sum_{q \leq Q} \mu^2(q) \prod_{\substack{p \mid q \\ p \neq 2}} \frac{2}{p - 2}$

to which we apply Theorem 21.1 with $\kappa = 2$ to get

(21.8) $G_1(Q) \sim \dfrac{1}{4} \displaystyle\prod_{p \geq 3} \dfrac{(p - 1)^2}{p(p - 2)} (\operatorname{Log} Q)^2$.

We again choose $Q = \sqrt{N}/\operatorname{Log} N$ to find that

$\left| \{ p \leq N \ / p + 2 \text{ is prime} \} \right| \leq 16(1 + o(1)) \displaystyle\prod_{p \geq 3} \dfrac{p(p - 2)}{(p - 1)^2} N/(\operatorname{Log} N)^2$

a bound that is 8 times larger than its conjectured value. (Siebert, 1976) establishes the above inequality for all $N > 1$ with no $o(1)$ term. If we were to use the Bombieri-Vinogradov Theorem as in section 13.5, we would get a bound only 4 times off the expected one. Note that (Wu, 2004) reduces this constant to 3.3996; that such an improvement holds only when we look at prime twins located on the initial segment $[1, N]$, contrarily to the above bound which remains valid for *any* interval of length N.

21.4. Products of four special primes in arithmetic progressions

Let us start by roughly recalling the notion of *sufficiently sifted sequence* as has been developed by (Ramaré & Ruzsa, 2001). Essentially, such a sequence $\mathcal{A}$ is infinite and of fairly large density: the number of its elements $\geq X$ is $\gg X/(\mathrm{Log}\,X)^{\kappa}$ for X large enough and a given κ; and for each large parameter X, we can find a $Y_X \leq X$, so that the finite subsequence $\mathcal{A} \cap [Y_X, X]$ can be sifted by a multiplicatively split compact set $\mathcal{K}$ satisfying the Johnsen-Gallagher condition, up to a level Q, in such a way that the associated G_1-function satisfies $G_1(Q) \gg (\mathrm{Log}\,X)^{\kappa}$. Alternatively, we may say that the characteristic function of $\mathcal{A} \cap [Y_X, X]$ is carried by $\mathcal{K}$ up to level Q. Such conditions ensure that the number of elements $\leq X$ in $\mathcal{A}$ is of order $X/(\mathrm{Log}\,X)^{\kappa}$ but also that we have at our disposal a surrounding compact. This latter condition provides us with good arithmetical properties: in (Ramaré & Ruzsa, 2001), we investigated its implications on additive properties; it is also a main ingredient in (Green & Tao, 2004) and (Green & Tao, 2006) concerning arithmetic progressions within such sets. We rapidly present here a third kind of use, namely to prove the existence of products of elements of this sequence in some arithmetic progressions to large moduli. This is, of course, a generalization of section 5.1.

But let us first comment some more on the definition of a *sufficiently sifted sequence* and provide the reader with some examples. The sequence of primes is a good candidate, with $\kappa = 1$. We see in this example that the introduction of Y_X is necessary: we cannot say that the primes up to X are the integers coprime with every integers $\leq Q = \sqrt{X}$ ... if we want to keep some elements in our sequence! Note that Q also has to depend on X, all of them conditions that gives a technical flavour to our definition but are required if we want it to be flexible enough for applications. A trivial example is also given by the sequence of positive integers, with $\kappa = 0$, or by the sequence of squarefree integers, also with dimension $\kappa = 0$. More exotic is the sequence of integers n that are sums of two coprime squares and such that $n + 1$ also shares this property. Its dimension is $\kappa = 1$, as shown in (Indlekofer, 1974/75). The sequence of those prime numbers p that can be written as $p = 1 + m^2 + n^2$ with $(m, n) = 1$ yields another uncommon example, with dimension $\kappa = \frac{3}{2}$ thanks to (Iwaniec, 1972).

Instead of going for a general result which would be very intricate, we use this latter sequence as an example: let $\mathcal{A}$ be the sequence of those prime numbers p that can be written as $p = 1+m^2+n^2$ with $(m,n) = 1$.

Theorem 21.3. *There exists $X_0 \geq 1$ and $h \geq 2$ with the following property. Let $X \geq X_0$ be an integer and $q_1,\dots, q_h$ be pairwise coprime moduli, all not more than $X^{1/3}$ and all prime to 3. Then modulo one of the q_i's, all invertible residue classes contain a product of four primes from $\mathcal{A}$, the four of them being not more than X.*

The bound $X^{1/3}$ may be replaced by $X^{\frac{1}{2}-\varepsilon}$ for any $\varepsilon > 0$ but then h may depend on ε. As a second remark, note that we detect a product of two primes, but in fact we can equally guarantee that each class modulo the same q_i contains also a product of five (or any number as well) primes from $\mathcal{A}$. Finally, we should mention that the modulus 3 is a special case since elements of $\mathcal{A}$ are congruent to 2 modulo 3 and in particular no product of a fixed number of them can cover all of $\mathcal{U}_3$.

The reader may try to get a similar result by taking for $\mathcal{A}$ the sequence of integers n and $n+1$ that are sums of two coprime squares. Note finally that (Pomerance *et al.*, 1988) somewhat draws on similar lines.

Proof. We split this proof in several steps.

General setting: We call $\mathcal{A}_X$ the sequence of elements of $\mathcal{A}$ that belong to $[\sqrt{X}, X]$, a sequence we can sieve up to level $Q = \sqrt{X}$. The cardinality of $\mathcal{A}_X$ is denoted by A_X.

The compact set $\mathcal{K}$ can be defined by multiplicativity: when $p \equiv 1[4]$, then $\mathcal{K}_p$ is simply the set of invertibles $\mathcal{U}_p$, while when $p \equiv 3[4]$, then $\mathcal{K}_p$ is the set of invertibles modulo p from which we remove the class 1. As for $p = 2$, we simply take $\mathcal{K}_2 = \{1\}$, without further ado. We then lift $\mathcal{K}_p$ trivially to define $\mathcal{K}_{p^\nu}$ for $\nu > 1$.

Applying Theorem 21.1, we find that $\kappa = 3/2$, from which we infer $G_1(Q) \gg (\text{Log } Q)^{3/2}$ while an appeal to Lemma 2.3 yields

$$(21.9) \qquad G_q(Q) \geq \acute{G}_1(Q/q) \gg (\text{Log}(Q/q))^{3/2},$$

the implied constant being of course independent of $q \leq Q$.

First step: By using (5.5), we get an analog of (5.2), namely that

$$G_1(Q)A_X^2 + \sum_{1 \leq i \leq h} G_{q_i}(Q)^{3/2}|K_{q_i}| \sum_{b \in \mathcal{K}_{q_i}} \left| \sum_{\substack{a \in \mathcal{A}_X \\ a \equiv b[q_i]}} 1 - A_X/|\mathcal{K}_{q_i}| \right|^2$$

is bounded from above by $A_X(X+Q^2)$. We then appeal to (21.9), recall that $Q = \sqrt{X}$ and get

$$\sum_{1\leq i\leq h} \left(1 - \frac{2\operatorname{Log} q_i}{\operatorname{Log} Q}\right)^{3/2} |K_{q_i}| \sum_{b\in K_{q_i}} \left|\frac{1}{A_X} \sum_{\substack{a\in A_X \\ a\equiv b[q_i]}} 1 - 1/|K_{q_i}|\right|^2 \leq c$$

for some constant $c > 0$ and all $X \geq X_0$. The introduction of this X_0 is necessary since we do not have $A_X \gg X/(\operatorname{Log} X)^{3/2}$ when X is too small. We use the same optimization process as in the proof of Theorem 5.1. First define

$$A_X(q_i) = \{a \in \mathbb{Z}/q_i\mathbb{Z}/ \ \exists p \in A_X, p \equiv a[q_i]\}.$$

Then there holds for $X \geq X_0$:

$$\sum_{1\leq i\leq h} \left(1 - \frac{2\operatorname{Log} q_i}{\operatorname{Log} X}\right)^{3/2} \left(\frac{|K_{q_i}|}{|A_X(q_i)|} - 1\right) \leq c.$$

From this inequality, we get, for one q_i we call q:

$$h \cdot (1/3)^{3/2} \left(\frac{|K_q|}{|A_X(q)|} - 1\right) \leq c$$

i.e. $|A_X(q)|/|K_q| \geq 1/(1 + 3c\sqrt{3}/h)$ which can be arbitrarily close to 1 when h is large enough.

Second step: We have just shown that the cardinality of $|A_X(q)|$ could be almost $|K_q|$ but this latter can be very small with respect to $\phi(q)$ when q has lots of prime factors $\equiv 3[4]$. However, we show here that the set $B(q)$ of products of two elements is large with respect to $\phi(q)$. The process we use to achieve this is rather classical. Let $r(n)$ (resp. $\tilde{r}(n)$) be the number of ways the integer n can be written as a product (resp. quotient) of two elements of $A_X(q)$ modulo q. Using Cauchy's inequality yields

$$|A_X(q)|^4 = \left(\sum_{n \bmod {}^*q} r(n)\right)^2 \leq |B(q)| \sum_{n \bmod {}^*q} r(n)^2.$$

We are to find an upper bound for $\sum_n r(n)^2$, but first we note that

$$\sum_{n \bmod {}^*q} r(n)^2 = \sum_{\substack{a,b,c,d\in A_X(q) \\ ab=cd[q]}} 1 = \sum_{\substack{a,b,c,d\in A_X(q) \\ a/c=d/b[q]}} 1 = \sum_{n \bmod {}^*q} \tilde{r}(n)^2.$$

Next, we compute an upper bound for $\tilde{r}(n)$ simply by extending in $n = a/b$ the range of a and b to all of K_q. This way, the new $\tilde{r}(n)$, say $\tilde{r}_0(q,n)$, is multiplicative. Furthermore, $\tilde{r}_0(p^\nu, n) = (|K_{p^\nu}|/|K_p|)\tilde{r}_0(p,n)$ for any $\nu \geq 1$. If $p \equiv 1[4]$, then $r_0(p,n) = |K_p| = p - 1$. Of course $\tilde{r}_0(2,n) = 1$. To cover the case $p \equiv 3[4]$, we note that in $a = nb$, all values of b are accepted when $n \equiv 1[p]$ and only $p - 3$ of them otherwise. From these remarks, and denoting by $q^{\sharp}$ the squarefree kernel of q, we get

$$\sum_{n \bmod^* q} \tilde{r}_0(q,n)^2 = \frac{|\mathcal{K}_q|^2}{|\mathcal{K}_{q^{\sharp}}|^2} \prod_{\substack{p|q, \\ p \equiv 1[4]}} ((p-1)|\mathcal{K}_p|^2)$$

$$\times \prod_{\substack{p|q, \\ p \equiv 3[4]}} ((p-2)(p-3)^2 + (p-2)^2)$$

$$= \frac{|\mathcal{K}_q|^4}{\phi(q)} \prod_{p|q, p \equiv 3[4]} \frac{(p-3)^2(p-1) + (p-2)(p-1)}{(p-2)^3}$$

$$= \frac{|\mathcal{K}_q|^4}{\phi(q)} \prod_{p|q, p \equiv 3[4]} \left(1 + \frac{1}{(p-2)^3}\right) \le 1.01 |\mathcal{K}_q|^4/\phi(q)$$

this latter inequality being true since the primes p intervening in the Euler product are ≥ 7. Gathering our estimates, we reach

$$(|\mathcal{A}_X(q)|/|\mathcal{K}_q|)^4 \le 1.01 |\mathcal{B}(q)|/\phi(q)$$

and thus for h large enough, we have $|\mathcal{B}(q)|/\phi(q) \ge 2/3$.

Third step. : We conclude as in the proof of Corollary 5.1: since $\mathcal{B}(q)$ contains more than $\phi(q)/2$ elements of $\mathcal{U}_q$, each class of $\mathcal{U}_q$ can be reached by a product of two elements from $\mathcal{B}(q)$.   ◇◇◇

Notations

Notations used throughout these notes are standard ... in one way or the other! Here is a guideline:

— The use of the letter p for a variable always implies this variable is a prime number.

— $e(y) = \exp(2i\pi y)$.

— $\Gamma(z)$ is the usual Euler Γ-function. In particular, $\Gamma(1/2) = \sqrt{\pi}$.

— $\|a\|_2$ stands for the norm, according to the ambient hermitian structure, or the L^2-norm when no such structure has been specified. This is to be distinguished from $\|u\|$ which stands for the distance to the nearest integer. In chapters 4 and 19, the norms will be denoted with another subscript, usually d or q, and it will still be hermitian norms and will *not* be linked in any way to L^q-spaces.

— $[d, d']$ stands for the lcm and (d, d') for the gcd of d and d'. We denote as usual the closed interval with endpoints M and N as $[M, N]$. Hermitian products will be denoted by $[f|g]$ with or without any subscript. And in chapter 20, we will denote by $[N/m]$ the integer part of N/m.

— $|\mathcal{A}|$ stands for the cardinality of the set $\mathcal{A}$ while $\mathbb{1}_{\mathcal{A}}$ stands for its characteristic function.

— $\mathbb{1}$ denotes a characteristic function in one way or another. For instance, $\mathbb{1}_{\mathcal{K}_d}$ is 1 if $n \in \mathcal{K}_d$ and 0 otherwise, but we could also write it as $\mathbb{1}_{n \in \mathcal{K}_d}$, closer to what is often called the Dirac δ-symbol. We shall also use $\mathbb{1}_{(n,d)=1}$ and $\mathbb{1}_{q=q'}$.

— $q\|d$ means that q divides d in such a way that q and d/q are coprime. In words we shall say that q *divides d exactly*.

— The squarefree kernel of the integer $d = \prod_i p_i^{\alpha_i}$ is $\prod_i p_i$, the product of all prime factors of d.

— $\omega(d)$ is the number of prime factors of d, counted without multiplicity.

— $\phi(d)$ is the Euler totient, i.e. the cardinality of the multiplicative group of $\mathbb{Z}/d\mathbb{Z}$.

— $\sigma(d)$ is the number of positive divisors of d, except in section 13.1 where it will denote a density.

— $\mu(d)$ is the Moebius function, that is 0 when d is divisible by a square > 1 and otherwise $(-1)^r$ otherwise, where r is the number of prime factors of d.

— $c_q(n)$ is the Ramanujan sum. It is the sum of $e(an/q)$ over all a modulo q that are prime to q. See also (8.12).

— $\Lambda(n)$ is van Mangoldt function: which is $\text{Log}\,p$ is n is a power of the prime p and 0 otherwise.

— The notation $f = \mathcal{O}_A(g)$ means that there exists a constant B such that $|f| \leq Bg$ but that this constant may depend on A. When we put in several parameters as subscripts, it simply means the implied constant depends on all of them.

— The notation $f = \mathcal{O}^*(g)$ means that $|f| \leq g$, that is a $\mathcal{O}$-like notation, but with an implied constant equal to 1.

— The notation $f \star g$ denotes the arithmetic convolution of f and g, that is to say the function h on positive integers such that $h(d) = \sum_{q|d} f(q)g(d/q)$.

— The notation $F * G$ denotes the real functions convolution, that is to say the function H on the real line defined by $H(x) = \int_{-\infty}^{\infty} F(x-y)G(y)dy$ provided the latter expression exists for every real number x.

— $\mathcal{U}$ is the compact set $(\mathcal{U}_d)_d$ where, for each d, $\mathcal{U}_d$ is the set of invertible elements modulo d.

— π is ... the usual real number about $3.141\,5\ldots$! But also identifies the counting function of the primes: $\pi(6) = 3$ for instance. We tried to avoid this notation when not too awkward, just as we did not use the Chebyshev ϑ and ψ functions except in chapter 6.

References

Abramowitz, M., & Stegun, I.A. 1964. *Handbook of mathematical functions.* Applied Mathematics Series, vol. 55. National Bureau of Standards. mintaka.sdsu.edu/faculty/wfw/ABRAMOWITZ-STEGUN.

Baier, S. 2006. On the large sieve with sparse sets of moduli. *J. Ramanujan Math. Soc.*, **21**(3), 279–295. Available at arxiv under reference math.NT/0512228.

Baier, S., & Zhao, L. 2005. Large sieve inequality with characters for powerful moduli. *Int. J. Number Theory*, **1**(2), 265–279.

Baier, S., & Zhao, L. 2006a. Bombieri-Vinogradov type theorems for sparse sets of moduli. *Acta Arith.*, **125**(2), 187–201. Available at arxiv under reference math.NT/0602116.

Baier, S., & Zhao, L. 2006b. Primes in Quadratic Progressions on Average. 13p. Available at arxiv under reference math.NT/0605563.

Barban, M.B. 1963. Über Analoga des Teilerproblems von Titchmarsh. *Vestn. Leningr. Univ.*, **18**(19 (Ser. Mat. Mekh. Astron. No.4)), 5–13.

Barban, M.B. 1964. Über die Verteilung der Primzahlen in arithmetischen Progressionen 'im Mittel'. *Dokl. Akad. Nauk UzSSR*, **5**, 5–7.

Barban, M.B. 1966. The 'large sieve' method and its application to number theory. *Uspehi-Math.-Naut*, **21**, 51–102. See also Russian Math. Surveys, 21 (1966), no 1, 49-103.

Basquin, J. 2006. Mémoire de DEA. *Université Lille 1*, 1–37.

Bateman, P.T. 1972. The distribution of values of the Euler function. *Acta Arith.*, **21**, 329–345.

Bertrandias, J.-P. 1966. Espaces de fonctions bornées et continues en moyenne asymptotique d'ordre *p*. *Mémoires de la Société Mathématique de France*, 3–106.

Boas, R.P.jun. 1941. A general moment problem. *Am. J. Math.*, **63**, 361–370.

Bombieri, E. 1965. On the large sieve method. *Mathematika*, **12**, 201–225.

Bombieri, E. 1971. A note on the large sieve method. *Acta Arith.*, **18**, 401–404.

Bombieri, E. 1976. The asymptotic sieve. *Rend., Accad. Naz. XL, V. Ser. 1-2*, 243–269.

Bombieri, E. 1987. Le grand crible dans la théorie analytique des nombres. *Astérisque*, **18**, 103pp.

Bombieri, E., & Davenport, H. 1966. Small differences between prime numbers. *Proc. Roy. Soc. Ser. A*, **293**, 1–18.

Bombieri, E., & Davenport, H. 1968. On the large sieve method. *Abh. aus Zahlentheorie und Analysis zur Erinnerung an Edmund Landau*, **Deut. Verlag Wiss., Berlin**, 11–22.

Bombieri, E., & Iwaniec, H. 1986. On the order of $\zeta(\frac{1}{2}+it)$. *Ann. Scuola Norm. Sup. Pisa Cl. Sci. (4)*, **13**(3), 449–472.

Bombieri, E., Friedlander, J.B., & Iwaniec, H. 1986. Primes in arithmetic progressions to large moduli. *Acta Math.*, **156**, 203–251.

Brüdern, J. 2000-2004. An elementary harmonic analysis of arithmetical functions.

Brüdern, J., & Perelli, A. 1999. Exponential Sums and Additive Problems Involving Square-free Numbers. *Ann. Scuola Norm. Sup. Pisa Cl. Sci.*, 591–613.

Cai, Yingchun, & Lu, Minggao. 2003. On the upper bound for $\pi_2(x)$. *Acta Arith.*, **110**(3), 275–298.

Cazaran, J., & Moree, P. 1999. On a claim of Ramanujan in his first letter to Hardy. *Expositiones Mathematicae*, **17**, 289–312. based on a lecture given 01-12-1997 by J. Cazaran at the Hardy symposium in Sydney.

Chen, Jing-run. 1978. On the Goldbach's problem and the sieve methods. *Sci. Sin.*, **21**, 701–739.

Cook, R.J. 1984. An effective seven cube theorem. *Bull. Aust. Math. Soc.*, **30**, 381–385.

Coppola, G., & Salerno, S. 2004. On the symmetry of the divisor function in almost all short intervals. *Acta Arith.*, **113**(2).

Coquet, J., Kamae, T., & Mendès France, M. 1977. Sur la mesure spectrale de certaines suites arithmétiques. *Bull. S.M.F.*, **105**, 369–384.

Croot III, E.S., & Elsholtz, C. 2004. On variants of the larger sieve. *Acta Math. Hung.*, **103**(3), 243–254.

Davenport, H. 1937a. On some infinite series involving arithmetical functions. *Quart. J. Math., Oxf. Ser.*, **8**, 8–13.

Davenport, H. 1937b. On some infinite series involving arithmetical functions. II. *Quart. J. Math., Oxf. Ser.*, **8**, 313–320.

Davenport, H., & Halberstam, H. 1966a. Primes in arithmetic progressions. *Michigan Math. J.*, **13**, 485–489.

Davenport, H., & Halberstam, H. 1966b. The values of a trigonometrical polynomial at well spaced points. *Mathematika*, **13**, 91–96.

Davenport, H., & Halberstam, H. 1968. Corrigendum: "Primes in arithmetic progressions". *Michigan Math. J.*, **15**, 505.

Duke, W., & Iwaniec, H. 1992. Estimates for coefficients of L-functions. II. *Pages 71–82 of:* E., Bombieri (ed), *Proceedings of the Amalfi conference on analytic number theory.*

Dusart, P. 1998. *Autour de la fonction qui compte le nombre de nombres premiers.* Ph.D. thesis, Limoges, http://www.unilim.fr/laco/theses/1998/T1998_01.pdf. 173 pp.

Elliott, P.D.T.A. 1971. Some remarks concerning the large sieve. *Acta Arith.*, **18**, 405–422.

Elliott, P.D.T.A. 1977. On the differences of additive arithmetic functions. *Mathematika*, **24**(2), 153–165.

Elliott, P.D.T.A. 1983. Subsequences of primes in residue classes to prime moduli. *Pages 507–515 of:* Erdös, P. (ed), *Studies in pure mathematics, Mem. of P. Turan.* Akadémia Kiadó, Budapest: Birkhäuser, Basel.

Elliott, P.D.T.A. 1985. Additive arithmetic functions on arithmetic progressions. *Proc. London Math. Soc.*, **54**(3), 15–37.

Elliott, P.D.T.A. 1985. *Arithmetic Functions and Integer Products.* Grundlehren der mathematischen Wissenschaften, vol. 272. Springer-Verlag New-York, Berlin, Heidelberg, Tokyo.

Elliott, P.D.T.A. 1991. On maximal variants of the Large Sieve. *J. Fac. Sci. Univ. Tokyo, Sect. IA*, **38**, 149–164.

Elliott, P.D.T.A. 1992. On maximal versions of the large sieve. II. *J. Fac. Sci. Univ. Tokyo, Sect. IA*, **39**(2), 379–383.

Elsholtz, C. 2001. The inverse Goldbach problem. *Mathematika*, **48**(1-2), 151–158.

Elsholtz, C. 2002. The distribution of sequences in residue classes. *Proc. Am. Math. Soc.*, **130**(8), 2247–2250.

Elsholtz, C. 2004. Upper bounds for prime k-tuples of size $\log N$ and oscillations. *Arch. Math.*, **82**(1), 33–39.

Erdös, P. 1937. On the sum and the difference of squares of primes. *J. Lond. Math. Soc.*, **12**, 133–136 and 168–171.

Erdös, P. 1940. The difference between consecutive primes. *Duke Math. J.*, **156**, 438–441.

Erdös, P. 1949. On a new method in elementary number theory which leads to an elementary proof of the prime number theorem. *Proc. Natl. Acad. Sci. USA*, **35**, 374–384.

Estermann, T. 1931. On the representations of a number as the sum of two numbers not divisible by a k-th power. *J. London Math. Soc.*, 37–40.

Estermann, T. 1938. On Goldbach's problem: Proof that almost all even positive integers are sums of two primes. *Proc. Lond. Math. Soc.*, II. Ser., **44**, 307–314.

Evelyn, C.J.A, & Linfoot, E.H. 1931. On a Problem in the additive Theory of Number (Third paper). *Mathematische Zeitschrift*, 637–644.

Forbes, T. Prime k-tuplets. *http://www.ltkz.demon.co.uk/ktuplets.htm*. From 1996.

Friedlander, J., & Iwaniec, H. 1992. A mean-value theorem for character sums. *Mich. Math. J.*, **39**(1), 153–159.

Friedlander, J., & Iwaniec, H. 1993. Estimates for character sums. *Proc. Am. Math. Soc.*, **119**(2), 365–372.

Friedlander, J.B., & Goldston, D.A. 1995. Some singular series averages and the distribution of Goldbach numbers in short intervals. *Illinois J. Math.*, **39**(1).

Gallagher, P.X. 1967. The large sieve. *Mathematika*, **14**, 14–20.

Gallagher, P.X. 1970. A large sieve density estimate near $\sigma = 1$. *Invent. Math.*, **11**, 329–339.

Gallagher, P.X. 1974. Sieving by prime powers. *Acta Arith.*, **24**, 491–497.

Goldfeld, D. 1985. Gauss's class number problem for imaginary quadratic fields. *Bull. Amer. Math. Soc. (1)*, **13**, 23–3?

Goldfeld, D.M., & Schinzel, A. 1975. On Siegel's zero. *Ann. Scuola Norm. Sup. Pisa Cl. Sci.*, **4**, 571–575.

Goldston, D.A. 1992. On Bombieri and Davenport's theorem concerning small gaps between primes. *Mathematika*, **1**.

Goldston, D.A. 1995. A lower bound for the second moment of primes in short intervals. *Expo. Math.*, **13**(4), 366–376.

Goldston, D.A., Pintz, J., & Yıldırım, C.Y. 2005. Primes in Tuples I. 36p. Available at arxiv under reference math.NT/0508185.

Graham, S.W., & Vaaler, J.D. 1981. A class of extremal functions for the Fourier transform. *Trans. Amer. Math. Soc.*, **265**(1), 283–302.

Granville, A., & Ramaré, O. 1996. Explicit bounds on exponential sums and the scarcity of squarefree binomial coefficients. *Mathematika*, **43**(1), 73–107.

Granville, A., & Soundararajan, K. 2003. Decay of mean values of multiplicative functions. *Can. J. Math.*, **55**(6), 1191–1230.

Greaves, G. 1974. An application of a theorem of Barban, Davenport and Halberstam. *Bull. London Math. Soc.*, **6**, 1–9.

Greaves, G. 2001. *Sieves in number theory*. Ergebnisse der Mathematik und ihrer Grenzgebiete, vol. 43. Springer-Verlag, Berlin. xii+304 pp.

Green, B., & Tao, T. 2004. The primes contain arbitrarily long arithmetic progressions. *Preprint*. Available at http://fr.arxiv.org/pdf/math.NT/0404188.

Green, B., & Tao, T. 2006. Restriction theory of the Selberg sieve, with applications. *J. Théor. Nombres Bordeaux*, **18**(1).

Gross, B., & Zagier, D. 1983. Points de Heegner et derivées de fonctions L. *C. R. Acad. Sci, Paris, Ser. I*, **297**, 85–87.

Gyan Prakash, & Ramana, D.S. 2008. Large inequality for integer polynomial amplitude. *J. Number Theory*. arXiv:0707.0671v1.

Halász, G. 1971/72. On the distribution of additive and the mean values of multiplicative arithmetic functions. *Studia Sci. Math. Hung.*, **6**, 211–233.

Halberstam, H., & Richert, H.E. 1971. Mean value theorems for a class of arithmetic functions. *Acta Arith.*, **43**, 243–256.

Halberstam, H., & Richert, H.E. 1974. Sieve methods. *Academic Press (London)*, 364pp.

Halberstam, H., & Richert, H.E. 1979. On a result of R. R. Hall. *J. Number Theory*, **11**, 76–89.

Hall, R.R. 1974. Halving an estimate obtained from Selberg's upper bound method. *Acta Arith.*, **25**, 347–351.

Hardy, G.H., & Littlewood, J.E. 1922. Some problems of "Partitio Numerorum" III. On the expression of a number as a sum of primes. *Acta Math.*, **44**, 1–70.

Heath-Brown, D.R. 1985. The ternary Goldbach problem. *Rev. Mat. Iberoamericana*, **1**(1), 45–59.

Heath-Brown, D.R. 1995. A mean value estimate for real character sums. *Acta Arith.*, **72**(3), 235–275.

Hensley, D., & Richards, I. 1974. Primes in intervals. *Acta Arith.*, **4**(25), 375–391.

Hildebrand, A. 1984. Über die punktweise Konvergenz von Ramanujan-Entwicklungen zahlentheoretischer Funktionen. *Acta Arith.*, **44**(2), 109–140.

Hoffstein, J. 1980. On the Siegel-Tatuzawa theorem. *Acta Arith.*, **38**, 167–174.

Holt, J.J., & Vaaler, J.D. 1996. The Beurling-Selberg extremal functions for a ball in euclidean space. *Duke Math. J.*, **83**(1), 202–248.

Huxley, M.N. 1968. The large sieve inequality for algebraic number fields. *Mathematika, Lond.*, **15**, 178–187.

Huxley, M.N. 1970. The large sieve inequality for algebraic number fields. II: Mean of moments of Hecke zeta-functions. *J. Lond. Math. Soc., III Ser.*, **21**, 108–128.

Huxley, M.N. 1971. The large sieve inequality for algebraic number fields. III: Zero-density results. *J. Lond. Math. Soc., II Ser.*, **3**, 233–240.

Huxley, M.N. 1972a. *The distribution of prime numbers. Large sieves and zero-density theorems.* Oxford Mathematical Monographs, Clarendon Press, Oxford. x+128 pp.

Huxley, M.N. 1972b. Irregularity in sifted sequences. *J. Number Theory*, **4**, 437–454.

Huxley, M.N. 1973. Small differences between consecutive primes. *Mathematika*, **20**, 229–232.

Indlekofer, K.H. 1974/75. Scharfe Abschätchung für die Anzahlfunction der *B*-Zwillinge. *Acta Arith.*, **26**, 207–212.

Iwaniec, H. 1972. Primes of the type $\varphi(x, y) + A$ where φ is a quadratic form. *Acta Arith.*, **21**, 203–234.

Iwaniec, H. 1980. Rosser's sieve. *Acta Arith.*, **36**, 171–202.

Iwaniec, H. 1994. *Analytic Number Theory.* Rutgers University. Lecture notes.

Iwaniec, H., & Kowalski, E. 2004. *Analytic number theory.* American Mathematical Society Colloquium Publications. American Mathematical Society, Providence, RI. xii+615 pp.

Ji, Chun-Gang, & Lu, Hong-Wen. 2004. Lower bound of real primitive *L*-function at $s = 1$. *Acta Arith.*, **111**(4), 405–409.

Johnsen, J. 1971. On the large sieve method in $GF[q, x]$. *Mathematika*, **18**, 172–184.

Kadiri, H. 2002. *Une région explicite sans zéros pour les fonctions L de Dirichlet.* Ph.D. thesis, Université Lille 1. tel.ccsd.cnrs.fr/documents/archives0/00/00/26/95/index_fr.html.

Kadiri, H. 2007. An explicit zero-free region for the Dirichlet *L*-functions. *To appear in J. Number Theory.*

Kobayashi, I. 1973. A note on the Selberg sieve and the large sieve. *Proc. Japan Acad.*, **49**(1), 1–5.

Konyagin, S.V. 2003. Problems of the set of square-free numbers. *Izv. Math.*, **68**(3), 493–520.

Landau, E. 1918. Über die Klassenzahl imaginär-quadratischer Zahlkörper. *Gött. Nachr.*, **1918**, 285–295.

Landau, E. 1935. Bemerkungen zum Heilbronnschen Satz. *Acta Arith.*, **1**, 1–18.

Levin, B.V., & Fainleib, A.S. 1967. Application of some integral equations to problems of number theory. *Russian Math. Surveys*, **22**, 119–204.

Linnik, Yu.V. 1941. The large sieve. *Doklady Akad. Nauk SSSR*, **30**, 292–294.

Linnik, Yu.V. 1942. A remark on the least quadratic non-residue. *Doklady Akad. Nauk SSSR*, **36**, 119–120.

Linnik, Yu.V. 1944a. On the least prime in an arithmetic progression. I: the basic theorem. *Mat. Sb., N. Ser.*, **15**(57), 139–178.

Linnik, Yu.V. 1944b. On the least prime in an arithmetic progression. II: the Deuring- Heilbronn theorem. *Mat. Sb., N. Ser.*, **15**(57), 139–178.

Linnik, Yu.V. 1961. The dispersion method in binary additive problems. *Leningrad*, 208pp.

Maier, H. 1988. Small differences between prime numbers. *Mich. Math. J.*, **35**(3), 323–344.

Martin, G. 2002. An asymptotic formula for the number of smooth values of a polynomial. *J. Number Theory*, **93**(2), 108–182.

McCurley, K.S. 1984. An effective seven cube theorem. *J. Number Theory*, **19**(2), 176–183.

Montgomery, H.L. 1968. A note on the large sieve. *J. London Math. Soc.*, **43**, 93–98.

Montgomery, H.L. 1971. Topics in Multiplicative Number Theory. *Lecture Notes in Mathematics (Berlin)*, **227**, 178pp.

Montgomery, H.L. 1978. The analytic principle of the large sieve. *Bull. Amer. Math. Soc.*, **84**(4), 547–567.

Montgomery, H.L. 1981. Maximal variants of the large sieve. *J. Fac. Sci., Univ. Tokyo, Sect. I A*, 805–812.

Montgomery, H.L., & Vaughan, R.C. 1973. The large sieve. *Mathematika*, **20**(2), 119–133.

Montgomery, H.L., & Vaughan, R.C. 2001. Mean values of multiplicative functions. *Period. Math. Hung.*, **43**(1-2), 199–214.

Motohashi, Y. 1977. A note on the large sieve. II. *Proc. Japan Acad. Ser. A Math. Sci.*, **53**(4), 122–124.

Motohashi, Y. 1978. Primes in arithmetic progressions. *Invent. Math.*, **44**(2), 163–178.

Motohashi, Y. 1979. A note on Siegel's zeros. *Proc. Jap. Acad., Ser. A*, **55**, 190–192.

Motohashi, Y. 1983. Sieve Methods and Prime Number Theory. *Tata Lectures Notes*, 205.

Oesterlé, J. 1985. Nombres de classes des corps quadratiques imaginaires. *Astérisque*, **121/122**, 309–323.

Pintz, J. 1976. Elementary methods in the theory of *L*-functions, II. *Acta Arith.*, **31**, 273–289.

Pomerance, C., Sárközy, A., & Stewart, C.L. 1988. On divisors of sums of integers. III. *Pacific J. Math.*, **133**(2), 363–379.

Preissmann, E. 1984. Sur une inégalité de Montgomery et Vaughan. *Enseign. Math.*, **30**, 95–113.

Puchta, J.-C. 2002. An additive property of almost periodic sets. *Acta Math. Hung.*, **97**(4), 323–331.

Puchta, J.-C. 2003. Primes in short arithmetic progressions. *Acta Arith.*, **106**(2), 143–149.

Ramachandra, K., Sankaranarayanan, A., & Srinivas, K. 1996. Ramanujan's lattice point problem, prime number theory and other remarks. *Hardy and ramanujan journal*, **19**.

Ramaré, O. 1995. On Snirel'man's constant. *Ann. Scu. Norm. Pisa*, **21**, 645–706.

Ramaré, O. 2005. Le théorème de Brun-Titchmarsh : une approche moderne. 1–10. http://math.univ-lille1.fr/~ramare/Maths/Nantes.pdf.

Ramaré, O. 2007a. Eigenvalues in the large sieve inequality. *Funct. Approximatio, Comment. Math.*, **37**, 7–35.

Ramaré, O. 2007b. An explicit result of the sum of seven cubes. *Manuscripta Math.*, **124**(1), 59–75.

Ramaré, O., & Ruzsa, I.M. 2001. Additive properties of dense subsets of sifted sequences. *J. Théorie N. Bordeaux*, **13**, 559–581.

Ramaré, O., & Schlage-Puchta, J.-C. 2008. Improving on the Brun-Titchmarsh Theorem. *Acta Arith.*, **131**, 351–366.

Rankin, R.A. 1947. The difference between consecutive prime numbers. III. *J. Lond. Math. Soc.*, **22**, 226–230.

Rankin, R.A. 1950. The difference between consecutive prime numbers. IV. *Proc. Am. Math. Soc.*, **1**, 143–150.

Rawsthorne, D.A. 1982. Selberg's sieve estimate with a one-sided hypothesis. *Acta Arith.*, **49**, 281–289.

Rényi, A. 1949. Un nouveau théorème concernant les fonctions independantes et ses applications à la théorie des nombres. *J. Math. Pures Appl.*, **IX Sér. 28**, 137–149.

Rényi, A. 1950. On the large sieve of Ju. V. Linnik. *Compos. Math.*, **8**, 68–75.

Rényi, A. 1958. On the probabilistic generalization of the large sieve of Linnik. *Magyar Tud. Akad. Mat. Kutató Int. Közl.*, **3**, 199–206.

Rényi, A. 1959. New version of the probabilistic generalization of the large sieve. *Acta Math. Acad. Sci. Hungar.*, **10**, 217–226.

Ricci, G. 1954. Sull'andamento della differenza di numeri primi consecutivi. *Riv. Mat. Univ. Parma*, **5**, 1–54.

Rosser, J.B., & Schoenfeld, L. 1962. Approximate formulas for some functions of prime numbers. *Illinois J. Math.*, **6**, 64–94.

Roth, K.F. 1964. Remark concerning integer sequences. *Acta Arith.*, **9**, 257–260.

Roth, K.F. 1965. On the large sieve of Linnik and Rényi. *Mathematika*, **12**, 1–9.

Schwartz, W., & Spilker, J. 1994. *Arithmetical functions, an introduction to elementary and analytic properties of arithmetic functions and to some of their almost-periodic properties.* Lectures Notes Series, vol. 184. Cambridge: London Math. Soc.

Selberg, A. 1942. On the zeros of the zeta-function of Riemann. *Norske Vid. Selsk. Forh., Trondhjem,* **15**(16), 59–62.

Selberg, A. 1943. On the zeros of Riemann's zeta-function. *Skr. Norske Vid.-Akad. Oslo,* **10**, 59 p.

Selberg, A. 1949a. An elementary proof of Dirichlet's theorem about primes in an arithmetic progression. *Ann. Math.,* **50**(2), 297–304.

Selberg, A. 1949b. An elementary proof of the prime-number theorem. *Ann. Math.,* **50**(2), 305–313.

Selberg, A. 1949. On elementary problems in prime number-theory and their limitations. *C.R. Onzième Congrès Math. Scandinaves, Trondheim, Johan Grundt Tanums Forlag,* 13–22.

Selberg, A. 1972 (August 14–18). Remarks on sieves. *Pages 205–216 of: Collected Works.* Number Theory Conference, University of Colorado, Boulder, Colorado.

Selberg, A. 1976. Remarks on multiplicative functions. *Lectures Notes in Mathematics (Berlin),* **626**, 232–241.

Selberg, A. 1991. Collected papers. *Springer-Verlag,* **II**, 251pp.

Siebert, H. 1976. Montgomery's weighted sieve for dimension two. *Monatsh. Math.,* **82**, 327–336.

Siegel, C.L. 1935. Über die Klassenzahl quadratischer Zahlkörper. *Acta Arith.,* **1**, 83–86.

Tatuzawa, T. 1951. On a theorem of Siegel. *Jap. J. of Math.,* **21**, 163–178.

Tchudakov, N.G. 1938. On the density of the set of even integers which are not representable as a sum of two odd primes. *Izv. Akad. Nauk SSSR,* **Ser. Mat. 1**, 25–40.

Uchiyama, Saburô. 1972. The maximal large sieve. *Hokkaido Math. J.,* **1**, 117–126.

Vaaler, J.D. 1985. Some Extremal Functions in Fourier Analysis. *Bull. A. M. S.,* **12**, 183–216.

van der Corput, J.G. 1937. Sur l'hypothèse de Goldbach pour presque tous les nombres pairs. *Acta Arith.,* **2**, 266–290.

van Lint, J.E., & Richert, H.E. 1965. On primes in arithmetic progressions. *Acta Arith.,* **11**, 209–216.

Vaughan, R.C. 1973. Some applications of Montgomery's sieve. *J. Number Theory,* **5**, 64–79.

Vaughan, R.C. 2003. Moments for primes in arithmetic progressions. *Duke Math. J.,* 371–383.

Vinogradov, A.I. 1965. The density hypothesis for Dirichet L-series. *Izv. Akad. Nauk SSSR Ser. Mat.*, **29**.

Vinogradov, I.M. 1937. Representation of an odd number as a sum of three primes. *Dokl. Akad. Nauk SSSR*, **15**, 291–294.

Wirsing, E. 1961. Das asymptotische Verhalten von Summen über multiplikative Funktionen. *Math. Ann.*, **143**, 75–102.

Wu, Jie. 1990. Sur la suite des nombres premiers jumeaux. *Acta Arith.*, **55**(4), 365–384.

Wu, Jie. 2004. Chen's double sieve, Goldbach's conjecture and the twin prime problem. *Acta Arith.*, **114**(3), 215–273.

Zhao, L. 2004a. Large sieve inequalities for special characters to prime square moduli. *Funct. Approximatio, Comment. Math.*, **32**, 99–106.

Zhao, L. 2004b. Large sieve inequality with characters to square moduli. *Acta Arith.*, **112**(3), 297–308.

Index